Advances in Anatomy, Embryology and Cell Biology publishes critical reviews and state-of-the-art surveys on all aspects of anatomy and of developmental, cellular and molecular biology, with a special emphasis on biomedical and translational topics.

The series publishes volumes in two different formats:

- Contributed volumes, each collecting 5 to 15 focused reviews written by leading experts
- Single-authored or multi-authored monographs, providing a comprehensive overview of their topic of research

226

Advances in Anatomy, Embryology and Cell Biology

More information about this series at
http://www.springer.com/series/102

Ennio Pannese

Biology and Pathology of Perineuronal Satellite Cells in Sensory Ganglia

Springer

Ennio Pannese
University of Milan
Milan
Italy

ISSN 0301-5556 ISSN 2192-7065 (electronic)
Advances in Anatomy, Embryology and Cell Biology
ISBN 978-3-319-60139-7 ISBN 978-3-319-60140-3 (eBook)
https://doi.org/10.1007/978-3-319-60140-3

Library of Congress Control Number: 2017955700

Printed on acid-free paper

This Springer imprint is published by the registered company Springer International Publishing AG part of Springer Nature
The registered company address is: Gewerbestrasse 11, 6330 Cham, Switzerland

To Marisa and Lucia

Acknowledgments

The author thanks Prof. F. Scaravilli (Department of Neuropathology, UCL, London, UK) and Prof. L. Zerboni (Department of Pediatrics/Infectious Diseases, Stanford University, Stanford, California, USA) who kindly provided micrographs, Mrs. A. Cerri for her advice during the making of the book, Mr. D. Ward for his help with the English, and Mr. G. Macor (Studio Macor, Milano) who digitalized the figures.

Contents

Abbreviations

CNS	Central nervous system
E face (of the plasma membrane)	Inner face of the outer (=External) leaflet of the split plasma membrane
kD	Kilodalton
mV	Millivolt
MΩ	Megohm
nm	Nanometer
P face (of the plasma membrane)	Outer face of the inner (=Protoplasmic) leaflet of the split plasma membrane
PNS	Peripheral nervous system
RNA	Ribonucleic acid
μm	Micrometer

Biology and Pathology of Perineuronal Satellite Cells in Sensory Ganglia

1.1 Historical Note

A cell sheath enveloping the body of the neurons in sensory ganglia was mentioned for the first time in 1836 by Valentin, a pupil of Purkinje. In some illustrations of his paper, the nuclei of cells adjacent to the surface of the nerve cell body, both in the trigeminal ganglion and in the ganglia of the autonomic nervous system, were clearly shown (Fig. 1.1a) even though they were misinterpreted as pigment granules. Since Remak (1838) denied the existence of this perineuronal sheath, Valentin (1839) provided a more detailed description of it, illustrated with new drawings (Fig. 1.1b), the captions of which gave a correct interpretation of the satellite cell nuclei.

Small perineuronal cells (or perineuronal nuclei) were then described in both sensory and autonomic ganglia by many authors under a variety of names such as *Polarkerne* (Courvoisier 1868), *Mantelzellen* and *Amphicyten* (Lenhossék 1897), *intrakapsuläre Zellen* (Holmgren 1901, 1902), *Randzellen* and *Scheidenzellen* (Kohn 1907), *Satellitenkörperchen* (Ramón y Cajal 1907) or *corpuscules ou cellules satellites* (Ramón y Cajal 1909), *Trophozyten* (Nemiloff 1908), *gliocitos* (De Castro 1921, 1946), *Hüllzellen, Hüllplasmodium* and *Nebenzellenplasmodium* (Stöhr jr. 1928), subcapsular cells (Penfield 1932), *Scheidenplasmodium* (Riegele 1932), capsular cells (Quade 1939), *ganglionar neuroglia* (Río Hortega et al. 1942), capsule cells (Kuntz and Sulkin 1947), and *periphere Glia* (Herzog 1954). Eventually, the term satellite cells became widely accepted. Since the term satellite cell is also used to refer to progenitor cells of skeletal muscle fibers, the satellite cells of sensory and autonomic ganglia are now often called satellite glial cells. However, oligodendrocytes and less frequently astrocytes may be found applied to the surface of nerve cell bodies in the CNS, so that the term satellite glial cells may refer to these oligodendrocytes and astrocytes. Therefore, to avoid confusion, the site of these cells should always be specified (e.g., satellite [glial] cells in the trigeminal ganglion).

E. Pannese, *Biology and Pathology of Perineuronal Satellite Cells in Sensory Ganglia*, Advances in Anatomy, Embryology and Cell Biology 226,
https://doi.org/10.1007/978-3-319-60140-3_1

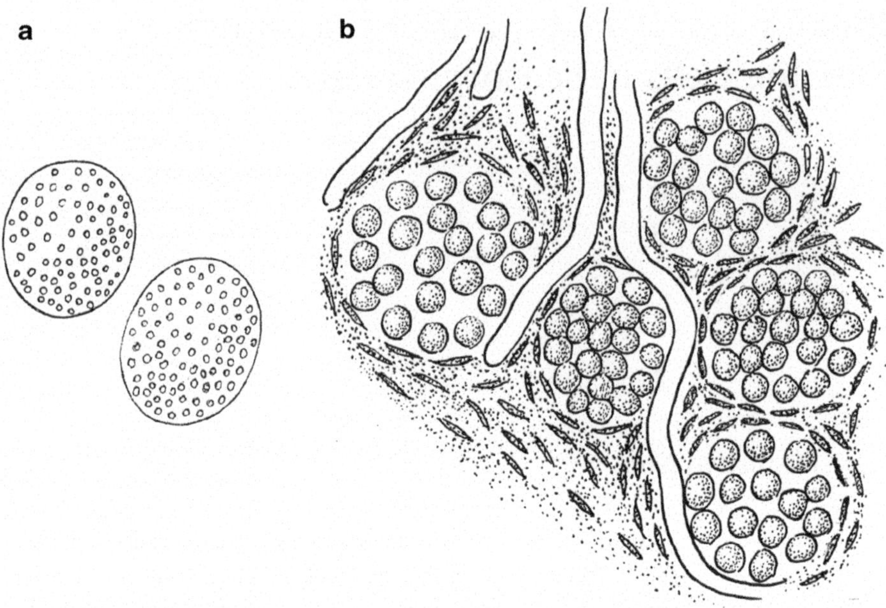

Fig. 1.1 Nerve cell bodies of sympathetic ganglia with the nuclei of satellite cells on the neuronal surface. Redrawn from Valentin; (**a**) 1836; (**b**) 1839. In his 1836 paper (**a**), Valentin misinterpreted these nuclei as pigment granules (*Pigmentkörperchen*), but in his 1839 paper (**b**) he gave a correct interpretation of the same nuclei

1.2 Organization of the Perineuronal Sheath in Adult Vertebrates

Suggestions that the perineuronal sheath might be absent in certain sensory ganglia or around some sensory neurons had been advanced in the past (e.g., see Schultze 1871; Lenhossék 1907; Levi 1907; Truex 1939). In light microscope preparations, some nerve cell bodies seem to be without a satellite cell envelope (Fig. 1.2). However, all the sensory neurons studied under the electron microscope have been shown to possess a satellite cell sheath. It was also debated whether the perineuronal sheath had a syncytial structure or consisted of discrete cells (for more details, see Pannese 1981). The advent of the electron microscope resolved this dispute showing that the sheath is built of discrete cells (Wyburn 1958 and many later authors).

In the sensory ganglia of adult vertebrates, both the nerve cell body and the initial segment of its axon are tightly enveloped by a satellite cell sheath, which is covered by a basal lamina. The satellite cell sheath with its basal lamina is in turn completely surrounded by the interstitial connective tissue of the ganglion (Fig. 1.3). Each nerve cell body together with its satellite cell sheath forms therefore a discrete structural unit (Pannese 1981). Furthermore, because of their close functional relationships (see Sects. 1.16 and 1.17), the body of the sensory ganglion neuron and its associated satellite cells also form a functional unit. Usually, each nerve cell body is completely isolated from every other; however, clusters of two or three nerve cell

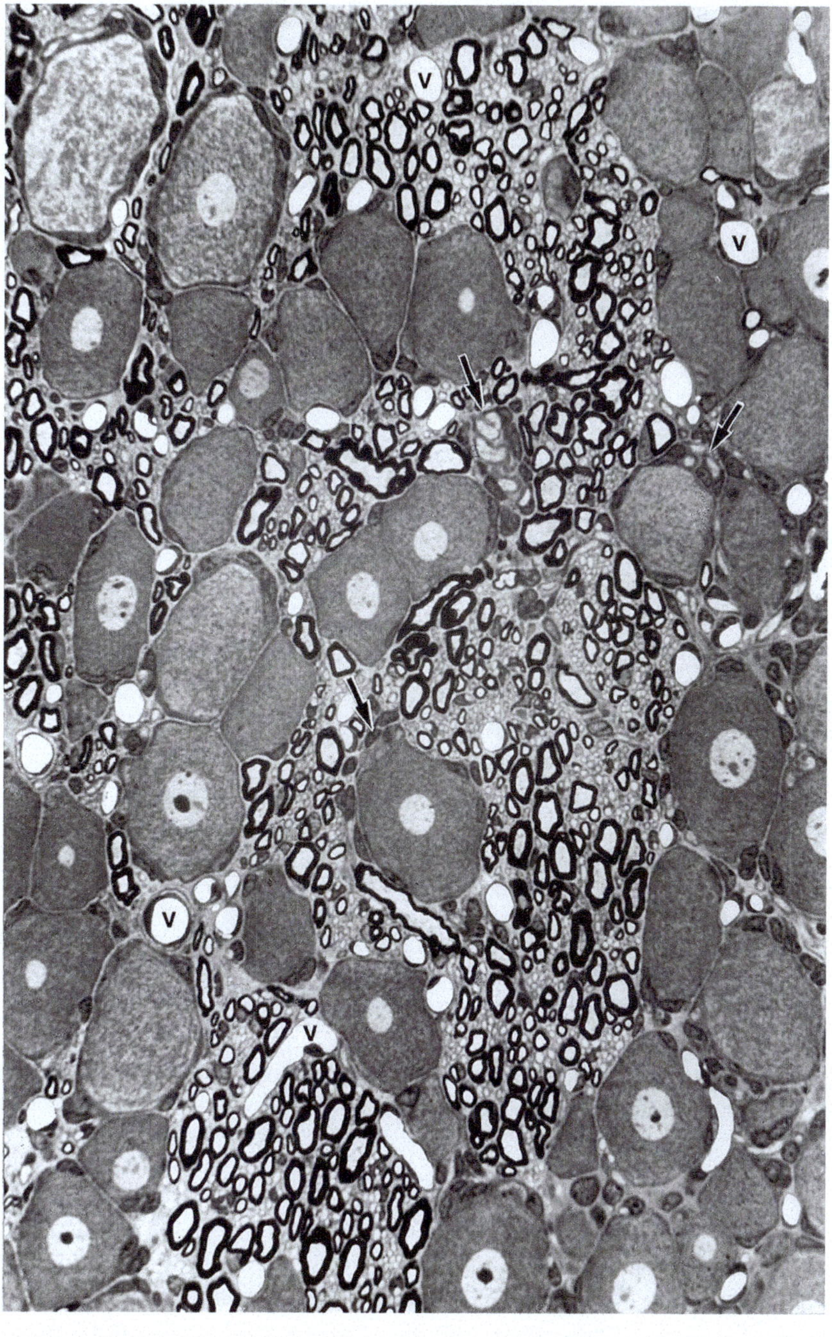

Fig. 1.2 Nerve cell bodies with their satellite cells (rabbit spinal ganglion). The nuclei of the satellite cells are evident, while their cytoplasm can be identified only in some places; consequently some portions of the neuronal surface seem to be without a satellite cell covering. *Arrows* point to axonal glomeruli. Note that there are numerous blood vessels; some are marked with **v**. Light micrograph of an 1-μm plastic section stained with toluidine blue, ×540

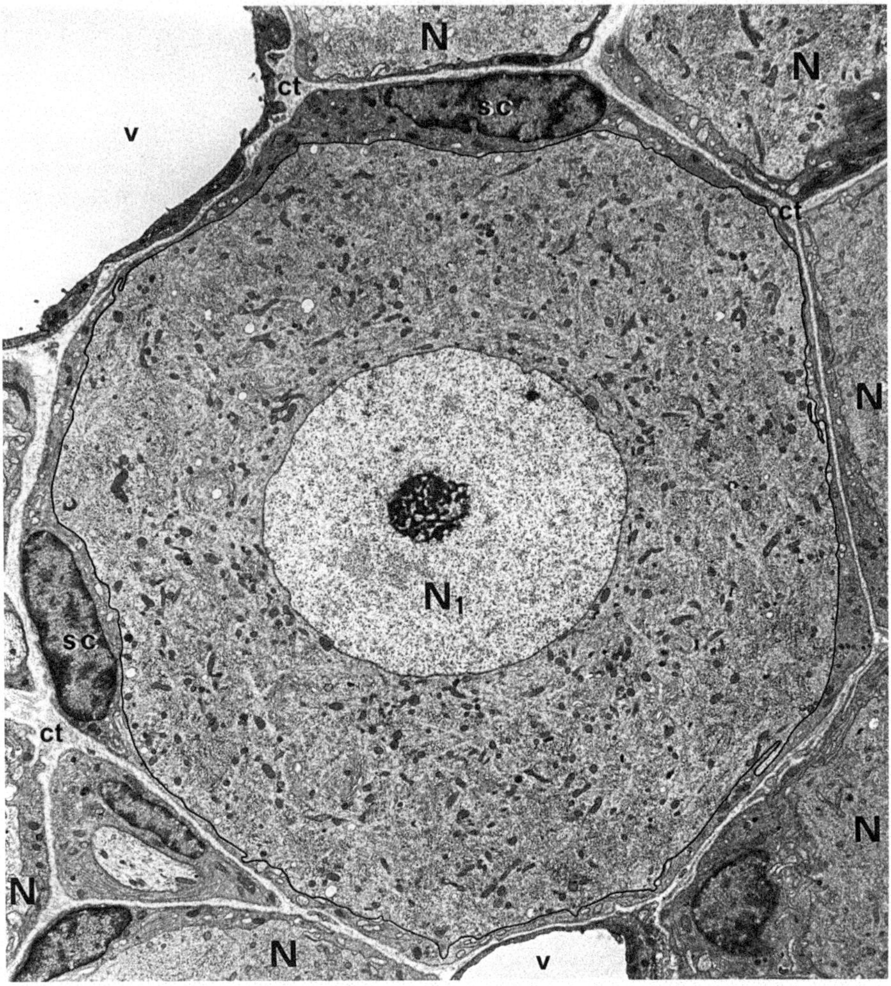

Fig. 1.3 Unit consisting of a nerve cell body (**N**$_1$) and its satellite cell (**sc**) sheath (rabbit spinal ganglion). The satellite cell sheath belonging to this unit is sharply separated from those encircling the adjacent nerve cell bodies (**N**) by the interstitial connective tissue (**ct**). The neuron-satellite cell boundary is outlined in ink. Note that the chromatin is distributed differently in the nuclei of satellite cells and in that of the neuron. **v** blood vessel. Transmission electron micrograph, ×4500

bodies enclosed within a common connective envelope have also been described (Fraentzel 1867; Holmgren 1901; De Castro 1921; Nawzatzky 1933; Mannu 1935; Hossack and Wyburn 1954; Wyburn 1958; McCracken and Dow 1973a; Pannese 1981). The percentage of nerve cell bodies arranged in clusters is highest in neonates and decreases progressively with age (Pannese et al. 1993). In the sensory ganglia of adult animals, the percentage of nerve cell bodies arranged in clusters range from 9.4 in the lizard, 5.6 in the rat, to 4.3 in the rabbit (Pannese et al. 1991, 1993). Sometimes, the nerve cell bodies of a given cluster are in direct mutual contact (Figs. 1.4 and 1.5); in other cases, they are separated from one another by a thin satellite cell sheet (Pannese et al. 1991). Interestingly, chemical transmission occurs

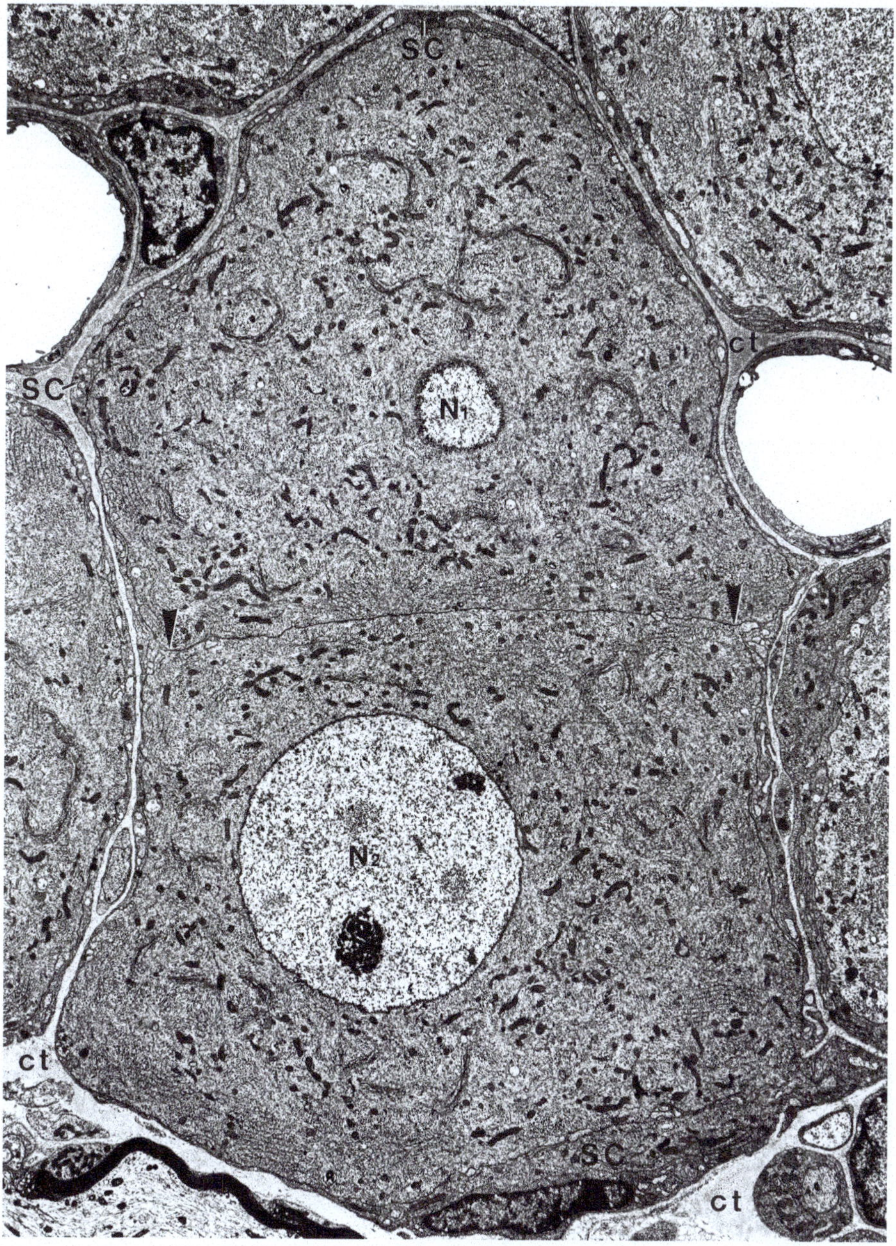

Fig. 1.4 A cluster of two nerve cell bodies (N_1 and N_2) enclosed within a common connective (**ct**) envelope (rat spinal ganglion). Between the two arrowheads, these nerve cell bodies are in direct contact. **SC** satellite cells. Transmission electron micrograph, ×4250

between the nerve cell bodies of a cluster even when the bodies themselves are separated by a thin satellite cell sheet (Rozanski et al. 2012, 2013).

The satellite cell sheath may show a variously complicated pattern of organization. It may consist of one single layer of flattened cells or of several layers of cells which

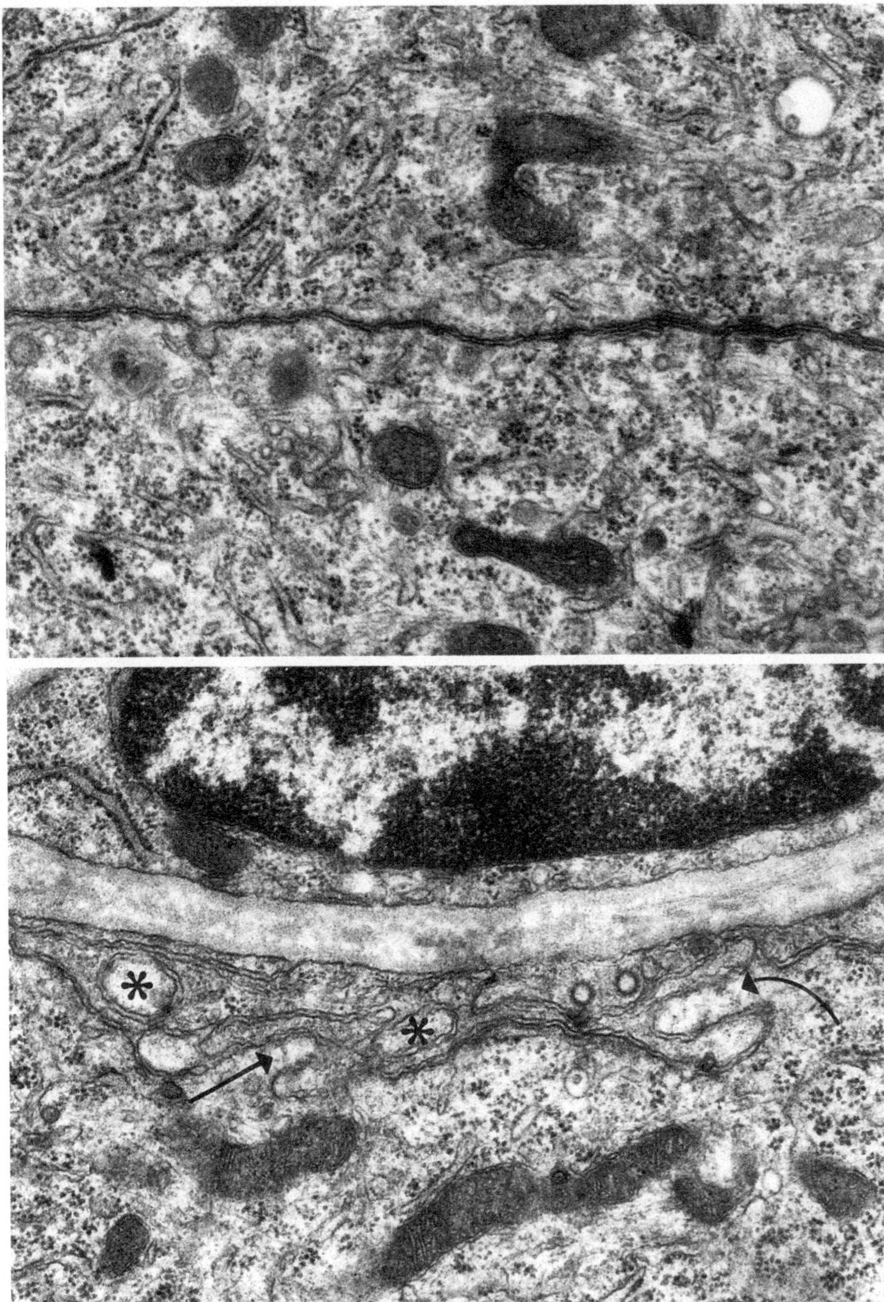

Fig. 1.5 Portions of the neurons of Fig. 1.4 are shown here at greater magnification (rat spinal ganglion). Perikaryal projections are absent from the neuronal surface in direct contact with the other neuron (*above*), but are present on the neuronal surface covered by satellite cells (*below*). *Arrows* point to projections which in this section appear in continuity with the neuronal perikaryon; * indicates projections which in this section appear as isolated entities because they arise from the neuronal perikaryon at other levels. Transmission electron micrograph, ×36,000

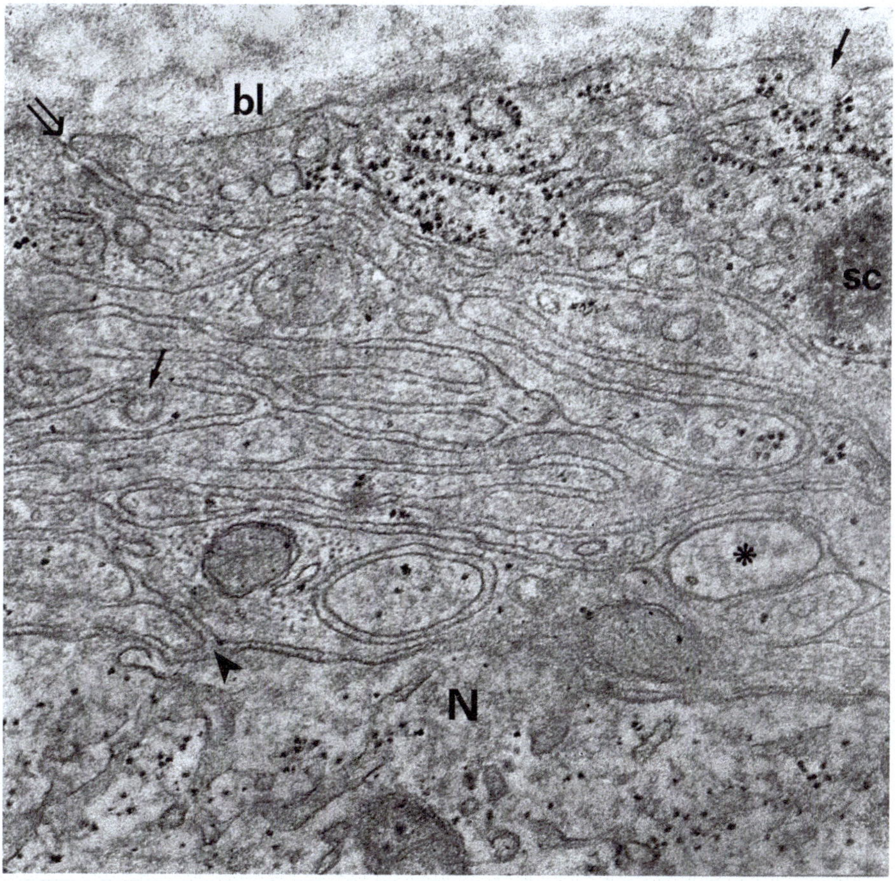

Fig. 1.6 Portion of a perineuronal sheath in which several layers of satellite cells overlap in a complex fashion (rabbit spinal ganglion). Cisternae of the granular endoplasmic reticulum are evident in the nucleated portion of a satellite cell; several vesicles can be seen as free in the satellite cell cytoplasm. *Arrows* point to pinocytotic vesicles; * indicates a cross section of a projection which arises from the neuronal perikaryon (**N**) at another level. **bl** basal lamina; **sc** satellite cell nucleus. Note that the clefts intervening between the apposed plasma membranes of adjacent satellite cells open into the light space under the basal lamina (*double arrow*) at one end and into the interval between the neuron and the satellite cell sheath at the other (*arrowhead*). Transmission electron micrograph, ×48,000

overlap in a complex fashion (Fig. 1.6). Both the arrangement of the satellite cells and the thickness of the sheath may vary from area to area within a given sheath. In spinal ganglia of *Xenopus laevis*, fowl, guinea pig, and rabbit, the sheath may vary in thickness from 40–50 nm to 4–6 μm; in the gecko it may vary from about 30 nm to about 3 μm and in the lizard from about 30 nm to about 1.5 μm. The sheath can sometimes be very thin, e.g., 18 nm in the eight cranial nerve ganglion of the goldfish (Rosenbluth and Palay 1961). Thus the satellite cell sheath can sometimes be below the limit of resolution of the light microscope. This explains why the sheath may have appeared discontinuous to some optical microscopists (e.g., Holmgren 1901; Penta 1934).

Although in electron microscope preparations, the satellite cell sheath usually appears as a continuous structure, gaps of various width may leave the surface of the neuron directly exposed to the basal lamina. Such gaps were occasionally observed in spinal ganglia of rat (Cervós-Navarro 1960; Andres 1961), bovine (McCracken and Dow 1973a), fowl, guinea pig (Pannese, unpublished observations), as well as in the spiral ganglion of the cat (Adamo and Daigneault 1973). In rabbit spinal ganglia, the gaps are from 0.06 to 0.75 μm in length (Pannese et al. 1996). The portions of the neuronal plasma membrane not covered by satellite cells are undercoated by a layer of dense material (Pannese et al. 1996), which is usually absent beneath the plasma membrane covered by satellite cells. These gaps are less rare in autonomic than in sensory ganglia and are more numerous and extensive in old age (see Sect. 1.15).

1.3 Shape of Satellite Cells

In the past, various opinions were expressed on the shape of satellite cells. (1) Satellite cells had a laminar shape (Fig. 1.7) and lacked processes (Hannover 1840, 1844; Fraentzel 1867; Key and Retzius 1873; Penta 1934; Palumbi 1944). (2) Satellite cells bore variably long and branching processes (Holmgren 1901, 1902; Ortiz-Picón 1932, 1949, 1955; Bertrand and Guillain 1933; Della Pietra 1937; Río Hortega et al. 1942; Scharenberg 1952; Stramignoni 1953; Herzog 1954; Fig. 1.8). Some of the satellite cells associated with the initial segment of the axon appeared as narrow sheets winding spirally around the initial segment (called *espirocitos* by Río Hortega). (3) Satellite cells lying close to the neuron bore processes while laminar cells forming an outer sheet lacked processes (Ramón y Cajal and Oloriz 1897; Ramón y Cajal 1909). The laminar cells were regarded by Ramón y Cajal (1909) as endothelial cells.

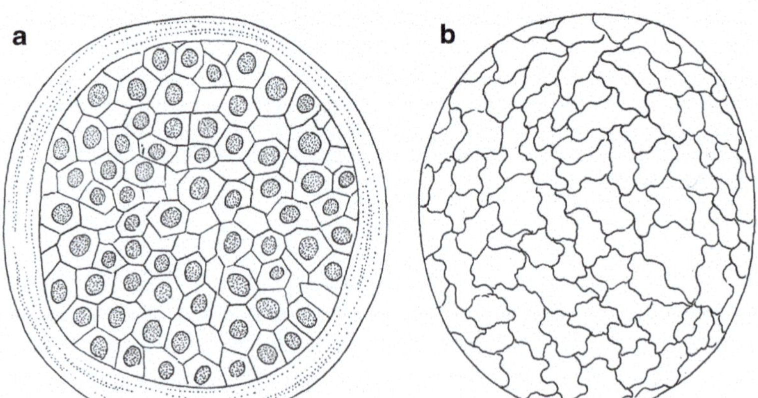

Fig. 1.7 The satellite cells are shown as laminar elements with an irregular polygonal shape. Redrawn and slightly modified from Fraentzel (1867). In (**a**) satellite cells are shown with their nuclei; in (**b**) only the outline of the same cells is evident

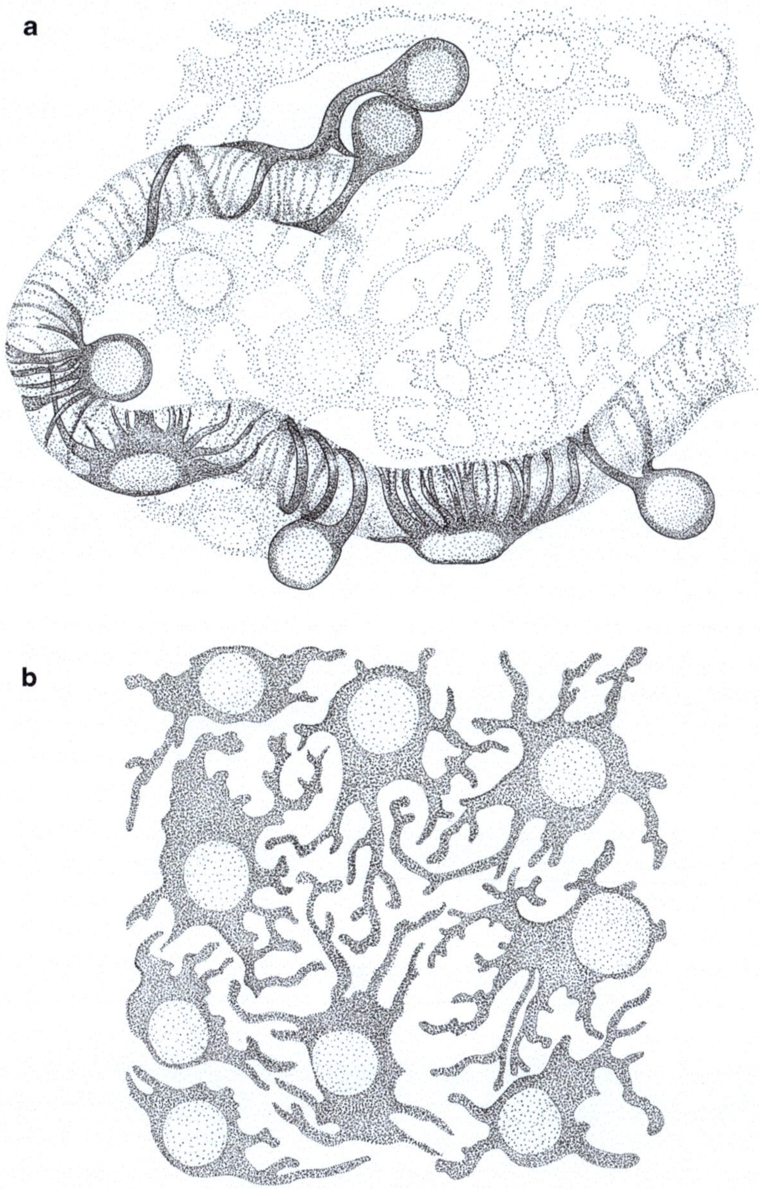

Fig. 1.8 According to the results of Río Hortega et al. (1942), both the satellite cells associated with the axon (**a**) and those associated with the nerve cell body (**b**) are shown as elements bearing processes

Experiments carried out on spinal ganglia using various fixatives have shown that the long, branching processes of satellite cells described in the past are the result of distortion of the cell shape (Fig. 1.9) due to the procedures used (Pannese 1960). As a rule, satellite cells are laminar and irregular in outline. Even after long-term in vitro culturing of sensory ganglia, satellite cells maintain shapes similar to

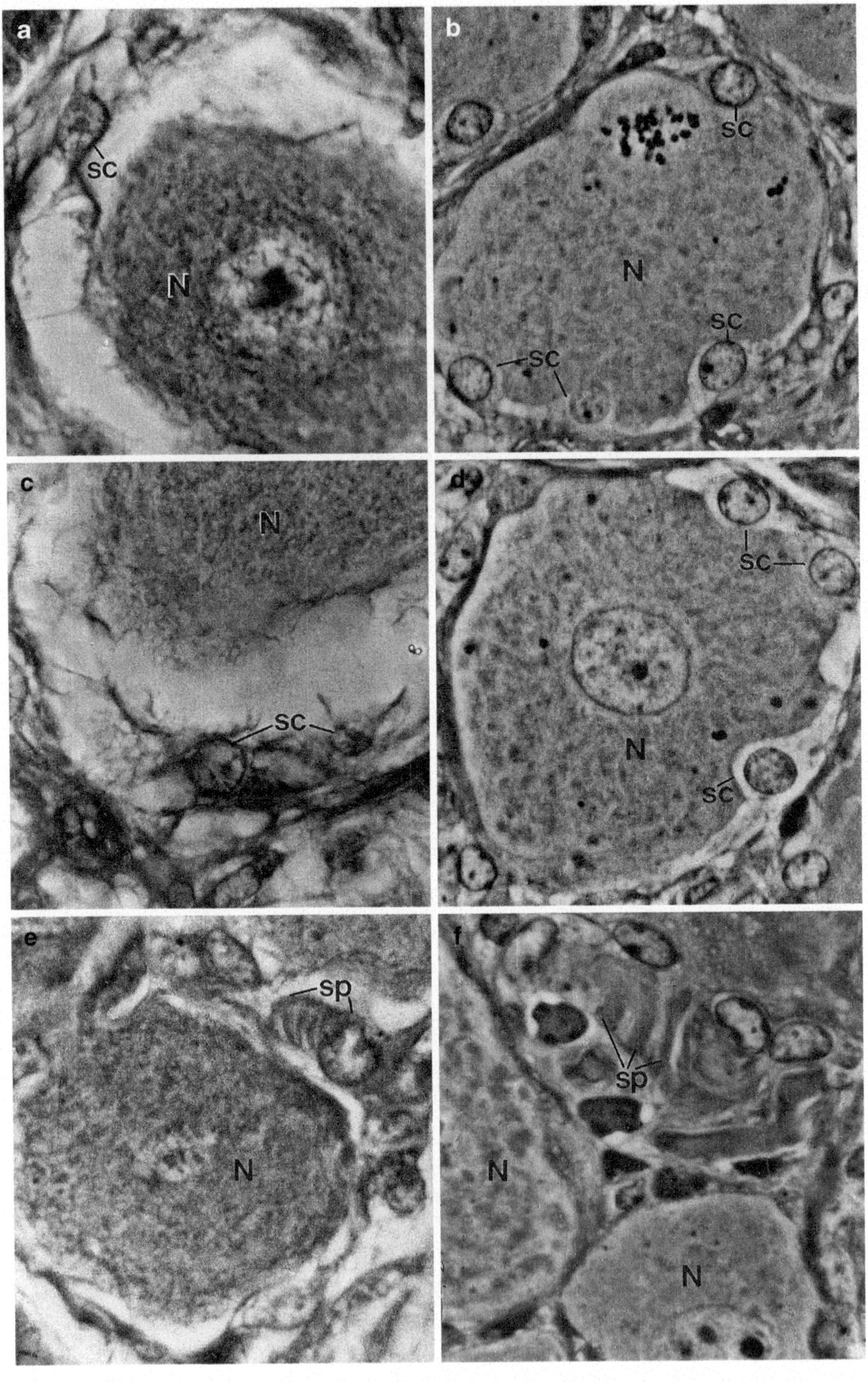
a
sc
N
b
sc
N
sc
sc
c
N
sc
d
sc
N
sc
e
sp
N
f
sp
N
N

those observed in vivo (Shimizu 1965; Bunge et al. 1967; Waxman et al. 1977). In cultures of dissociated ganglia, these cells often appear aggregated in an epithelial-like pattern. By 1873 Key and Retzius had already suggested that satellite cell processes could be artifactual formations due to shrinkage of the ganglionic neuron. Even the cleft which is sometimes apparent between the nerve cell and the perineuronal connective tissue in light microscope preparations (Fig. 1.9) is the result of an artifactual shrinkage of the neuronal body, as suggested by several early authors (e.g., Koneff 1887; Flemming 1895; Schaffer 1896; Lenhossék 1897; Buehler 1897; Nemiloff 1908).

Satellite cells are endowed with projections (Fig. 1.20) that increase their surface area. The projections may be fingerlike (from 50 to 100 nm in diameter) or may have a lamellar shape (from 30 to 100 nm in thickness). They are numerous especially at the cell surface facing the sensory neuron. It is not known whether these projections are stable or transient structures.

1.4 Structure of Satellite Cells

In general, the satellite cell has only one nucleus. Binucleate satellite cells have only occasionally been observed (Kubota and Hioki 1943; Forssmann 1964). In several species, the nucleated portion of the satellite cell is apposed to a flattened area of the neuronal surface; in the cat, on the other hand, the nucleated portion of the satellite cell indents the neuronal surface. The nucleus is often elliptical (Fig. 1.2), less frequently round, kidney shaped, or nearly triangular in section; in the South African toad, it often shows one or more deep indentations. Chromatin (Fig. 1.3) usually occurs in centrally located clumps and in a peripheral band (Bunge et al. 1967; Pineda et al. 1967), which is separated from the inner surface of the nuclear envelope by a fibrous lamina. In the nuclei of neurons ensheathed by satellite cells, by contrast, the chromatin is usually widely dispersed (Fig. 1.3).

Free polysomes and cisternae of the granular endoplasmic reticulum (Figs. 1.6 and 1.10) are common in the cytoplasm (Hess 1955; Wyburn 1958). These cisternae may be arranged in stacks (Pannese 1960, 1964, 1969; Bunge et al. 1967) [Fig. 1.11]. These stacks of cisternae with their associated ribosomes may correspond to the basophilic granules seen in satellite cells under the light microscope by Kubota and Hioki (1943) and Palumbi (1944). Subsurface cisternae (Fig. 1.12) may

Fig. 1.9 Shape of satellite cells in spinal ganglia fixed in Bouin's fluid (**a**, **c**), formaldehyde (**e**), and OsO_4 (**b**, **d**, and **f**). (**a**, **b**, and **d**) horse; (**c**) ox; (**e**) rabbit; (**f**) guinea pig. Note that in the ganglia fixed in Bouin's fluid (**a**, **c**), a cleft is apparent around the nerve cell body (**N**) where satellite cells (**sc**) bearing branching processes can be seen. In OsO_4-fixed ganglia (**b**, **d**, and **f**), no cleft can be seen around the nerve cell body (**N**), and the satellite cells (**sc**) lie in close contact with the latter and appear as laminar elements. **sp** spiral processes of satellite cells coiled around axons. Phase contrast micrographs, (**a**–**d**) ×1000; (**e**, **f**) ×1450

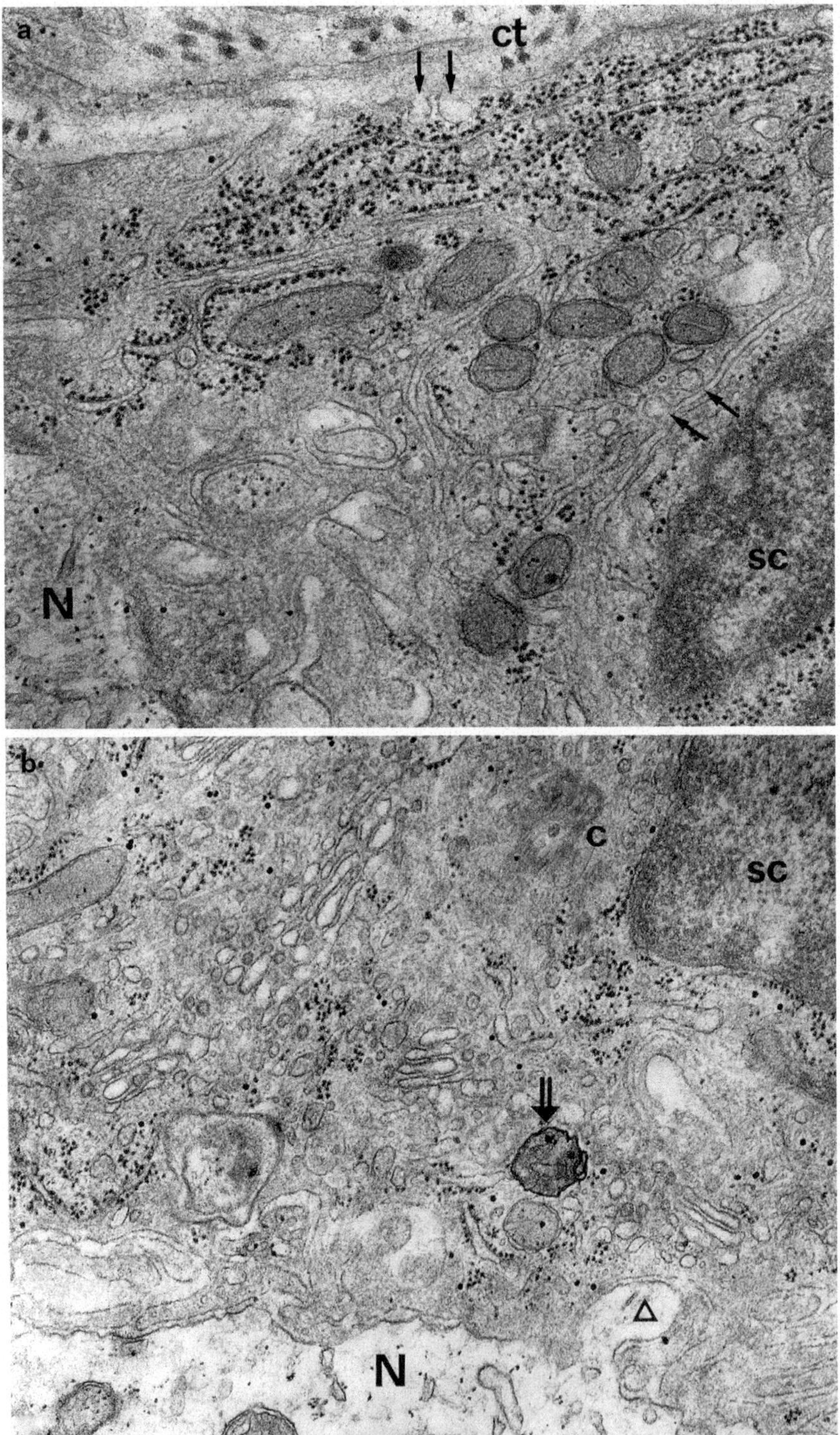
a
ct
N
sc
b
c
sc
N

be observed closely applied to the plasma membrane adjacent to another satellite cell and rarely to the plasma membrane facing the neuron (see also Sect. 1.8). Confronting subsurface cisternae can rarely be seen in two adjacent satellite cells. Tubules and cisternae of the agranular endoplasmic reticulum are less frequently observed, except in frog autonomic ganglia, where they are abundant. The Golgi apparatus usually lies near the nucleus. Under the electron microscope, each section of this organelle appears built of a stack of three to six flattened and smooth-surfaced cisternae with a cluster of associated vesicles (Fig. 1.10).

Mitochondria (Fig. 1.10) show the distinctive structural features well known from the literature (e.g., see Pannese 2015). Mitochondrial cristae, in general, take

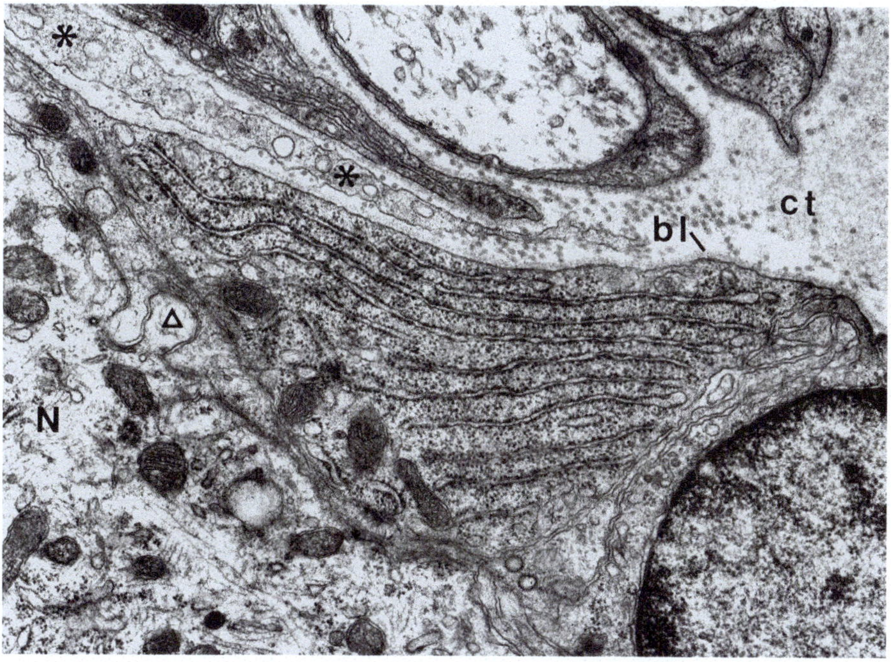

Fig. 1.11 Numerous cisternae of the granular endoplasmic reticulum arranged in a stack can be seen in the satellite cell cytoplasm (rat spinal ganglion). The outer contour of the satellite cell sheath facing the connective tissue (**ct**) is lined by a basal lamina (**bl**). Note that the cytoplasmic process of a connective tissue cell (*) is devoid of a basal lamina. *Open triangle* indicates a projection of the neuronal perikaryon (**N**) which invaginates the inner contour of the satellite cell sheath. Transmission electron micrograph, ×21,800

Fig. 1.10 Cytoplasmic organelles in satellite cells (rabbit spinal ganglion). Cisternae of the granular endoplasmic reticulum and mitochondria can be seen in (**a**); a centriole (**c**), an autophagic vacuole (*double arrow*), and several sections of the Golgi apparatus can be seen in (**b**). *Arrows* point to pinocytotic vesicles; *open triangle* indicates a neuronal projection which invaginates the inner contour of the satellite cell sheath. **ct** connective tissue; **N** neuronal perikaryon; **sc** satellite cell nucleus. Transmission electron micrograph, ×40,000

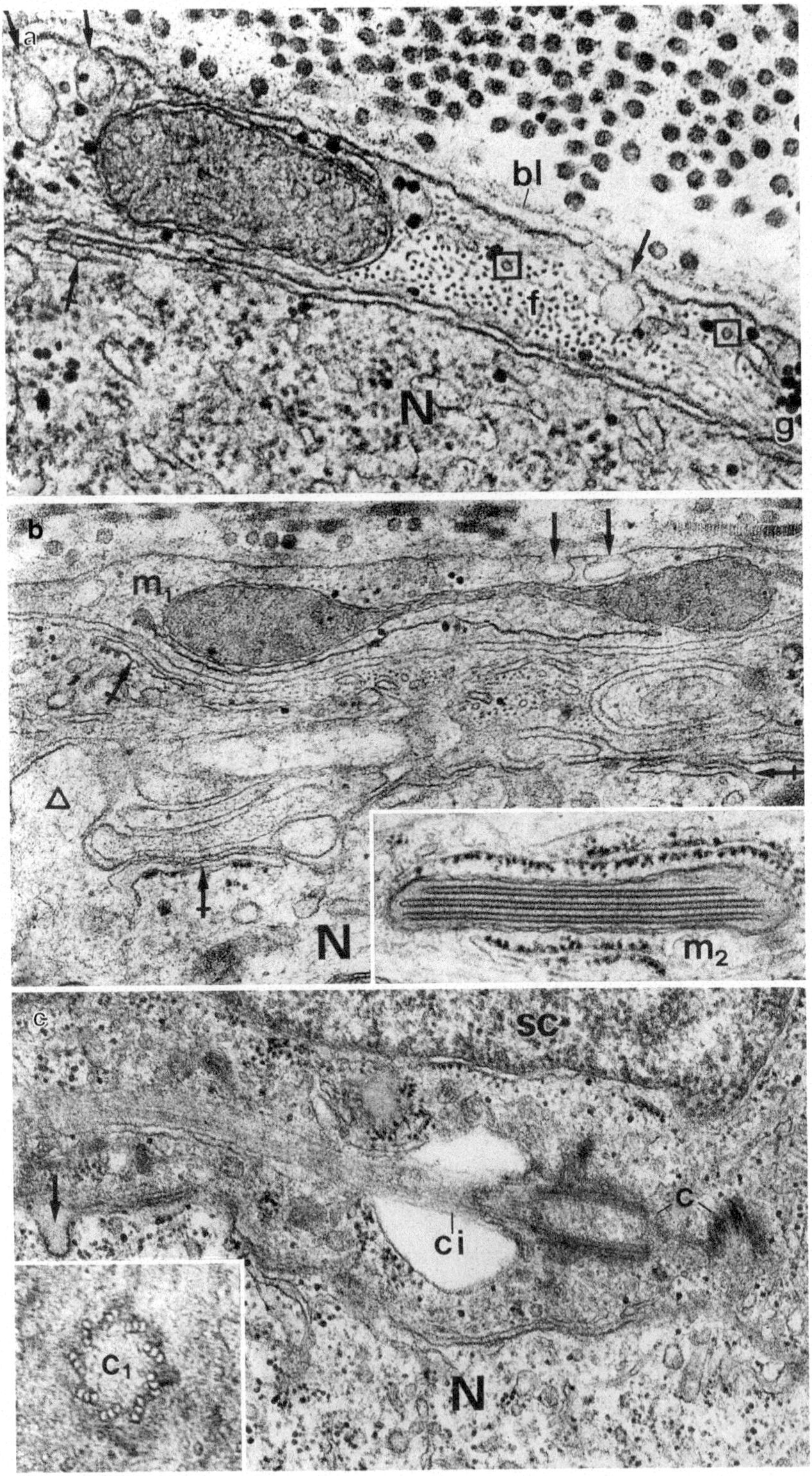
a
bl
f
N
g
b
m1
N
m2
c
sc
ci
c
c1
N

the form of thin folds and, less frequently, of slender tubules; both platelike and tubular cristae may be found in the same mitochondrion. Cristae are usually oriented transversely to the long axis of the mitochondrion; sometimes however they are oriented longitudinally (Fig. 1.21). Cristae oriented transversely at one end of the mitochondrion and longitudinally at the other end can also be observed. Small dense granules may be found in the mitochondrial matrix. In rabbit spinal ganglia, a small percentage of mitochondria contain, together with common cristae, atypical cristae, resembling those described by Newcomb et al. (1968) in mitochondria of plant cells. These cristae (Fig. 1.12), which always occur in elongated mitochondria, have a major axis approximately coincident with that of the mitochondrion; their membrane appears denser than that of the common cristae. Similar cristae have also been described in neuroglial cells of the CNS of lizard under apparently normal conditions (Gray 1960) and in other cell types under experimentally modified conditions (e.g., Adoutte et al. 1972; Tuchweber et al. 1972). In rabbit spinal ganglia, the mean percentage of cytoplasmic volume occupied by mitochondria is about 30% greater in nerve cell bodies than in their associated satellite cell sheaths (Martinelli et al. 2006b, 2007). Elongated organelles with two paired enveloping membranes and a filamentous content have been occasionally observed in satellite cells of the frog spinal ganglia (Kohno 1969). These bodies, whose significance is unclear, have been called gliosomes as they resemble organelles found in neuroglial cells of the CNS (Held 1909; Fieandt 1910 and, more recently, Farquhar and Hartmann 1957; Fleischhauer 1958; Gray 1959; Srebro 1965; Donelli et al. 1975) and named *gliosomen* (gliosomes) by Fieandt (1910). These organelles should not be confused with the glial plasmalemmal vesicles described by Nakamura et al. (1993), which have also been termed gliosomes.

Lysosomes have been detected cytochemically in spinal ganglia of rat (Citkowitz and Holtzman 1973) and chick embryo (Pannese et al., unpublished observations). Autophagic vacuoles (Fig. 1.10) are sometimes present in satellite cells. Whether they contain lytic enzymes and, therefore, are autolysosomes, has not yet been determined. Peroxisomes have been detected cytochemically in spinal ganglia of rat (Citkowitz and Holtzman 1973) and in the trigeminal ganglion of mouse (Hanker et al. 1974). They appear as small (about 0.2 μm in diameter), membrane-bounded bodies with a matrix of moderate electron density but no core or nucleoid. Multivesicular bodies are also present (Pannese 1964, 1969; Bunge et al. 1967).

Microtubules (Bunge et al. 1967; Pannese 1969), 20–30 nm in diameter, and intermediate filaments (Rosenbluth 1962a; Pannese 1964; Moses et al. 1965), about

Fig. 1.12 Cytoplasmic organelles in satellite cells (spinal ganglia of bullfrog (**a**), rabbit (**b**), and lizard (**c**)). *Arrows* point to pinocytotic vesicles; *crossed arrows* point to subsurface cisternae; *open triangle* indicates a neuronal projection which invaginates the inner contour of the satellite cell sheath; *white square* encloses cross-sectioned microtubules; **bl** basal lamina; **c** centrioles, one of which serves as the basal body of a cilium (**ci**); $\mathbf{c_1}$ a centriole in cross section; **f** cross-sectioned filaments; **g** glycogen particles; $\mathbf{m_1}$ dumbbell-shaped mitochondrion; $\mathbf{m_2}$ elongated mitochondrion with longitudinally oriented "atypical" cristae; **N** neuronal perikaryon; **sc** satellite cell nucleus. Transmission electron micrographs. (**a**) ×64,200; (**b**) and inset ×42,800; (**c**) ×32,100; and inset ×64,200

10 nm in thickness, occur throughout the cytoplasm, generally lying more or less parallel to both the outer and inner surfaces of the satellite sheath (Figs. 1.12 and 1.21). They are particularly numerous around the axon hillock and the initial segment of the axon.

Centrioles, observed by Kubota and Hioki (1943) under the light microscope, usually lie near the nucleus and the Golgi complex (Figs. 1.10, 1.12 and 1.13). Under the electron microscope, they appear as cylinders (0.3–0.5 μm in length and 0.2 μm in diameter), whose wall (Fig. 1.12) is composed of the usual nine triplets of microtubules (Pannese 1964, 1969; Bunge et al. 1967). Solitary primary cilia (Figs. 1.12 and 1.13) have been occasionally observed in satellite cells (autonomic ganglia: Grillo and Palay 1963; sensory ganglia: Pannese 1964). The cilium originates from a basal body (Fig. 1.13), which sometimes bears coarse appendages extending radially from its wall and, more rarely, a cross-striated rootlet. The shaft of the cilium, 0.2–0.3 μm in diameter, projects into a narrow channel of the extracellular space formed by a deep invagination of the plasma membrane and may be directed toward the adjacent neuron or the connective tissue space. These cilia lack the central pair of microtubules and only show the nine pairs of peripheral microtubules, as do those in neurons, in Schwann cells, and in astrocytes. The finding of specific receptors on the solitary cilia of certain neurons (for a review, see Pazour and Witman 2003) supports the hypothesis that these cilia are sensory organelles. The cilia of satellite cells may have a similar sensory function; however, there appear to be no data on the presence of specific receptors on the cilia of these cells.

Lipofuscin granules may occur in the satellite cells of sensory ganglia (Andres 1961; Pannese 1964; Glees and Gopinath 1973). They are more often present in the perinuclear cytoplasm than elsewhere. These granules are bounded by a membrane and have a dense matrix which contains homogeneous vacuoles of lower density. Membranous arrays can sometimes be seen in the dense matrix. It has been suggested that satellite cells actively cooperate in the elimination of lipofuscin accumulated within spinal ganglion neurons. Lipofuscin granules would be transferred from the neuronal perikaryon to the cytoplasm of surrounding satellite cells and then to the connective tissue and, finally, removed via the capillaries (Glees and Gopinath 1973). However, observations in rat and rabbit spinal ganglia do not support this hypothesis (van den Bosch de Aguilar and Vanneste 1983; Ledda et al. 1999). Glycogen particles (25–30 nm in diameter) may occur singly or grouped in small clusters (Fig. 1.12). Lipid inclusions (Fig. 1.14) have been found in satellite cells (Kubota and Hioki 1943; Andres 1961; Rosenbluth 1962a; Cravioto and Merker 1963; Arnold 1970). In mammals and birds, these inclusions are usually scarce and do not exceed 1 μm in diameter; in amphibians, by contrast, at certain periods of the year, they can be very numerous and so large (several micrometers in

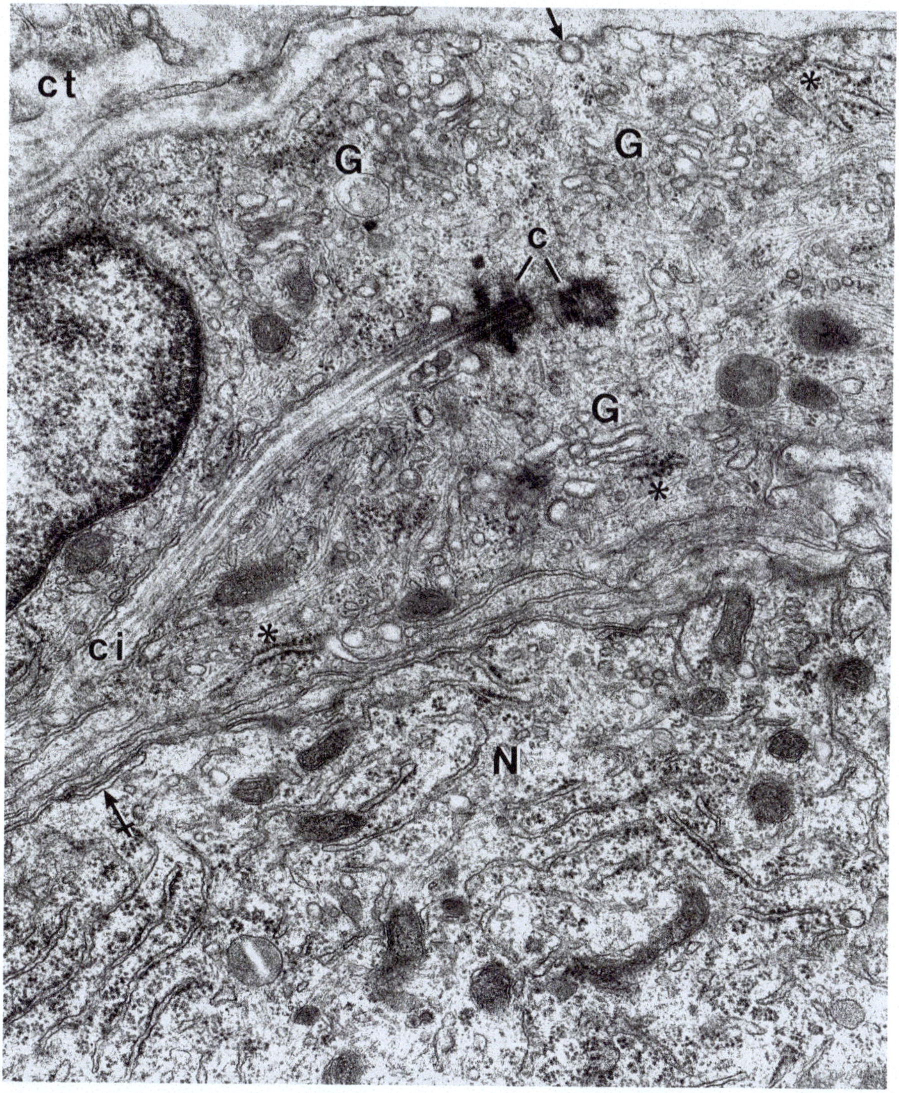

Fig. 1.13 Mitochondria, dense bodies, cisternae of the granular endoplasmic reticulum (*), and Golgi complexes (**G**) can be seen in the cytoplasm of a satellite cell. Microtubules and intermediate filaments are interposed between the other organelles (rabbit spinal ganglion). *Arrow* points to a pinocytotic vesicle along the satellite cell plasma membrane facing the interstitial connective tissue (**ct**). *Crossed arrow* points to a subsurface cisterna. **c** centrioles, one of which serves as the basal body of a cilium (**ci**); **N** neuronal perikaryon. Transmission electron micrograph, ×28,000

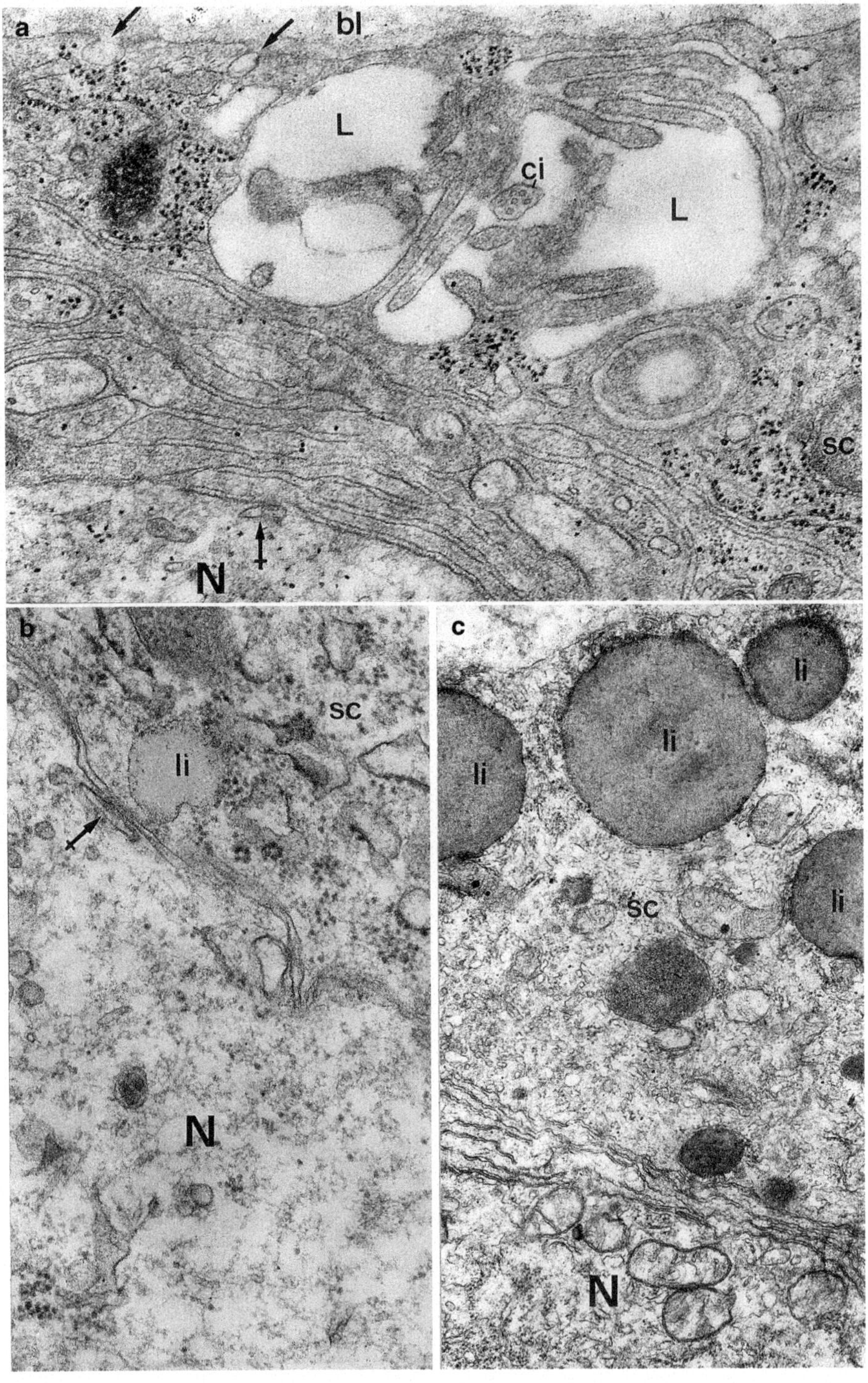
a
bl
L
ci
L
SC
N
b
SC
li
N
c
li
li
li
li
SC
N

Fig. 1.14 (**a**) Cytoplasmic lacuna (**L**) in a satellite cell (rabbit spinal ganglion). Some microvilli and a cilium (**ci**) project into the lacunar space. *Arrows* point to pinocytotic vesicles, and the *crossed arrow* points to a subsurface cisterna; **bl** basal lamina; **N** neuronal perikaryon; **sc** satellite cell nucleus. (**b, c**) Lipid inclusions (**li**) in satellite cells (**b** fowl spinal ganglion; **c** *Xenopus* spinal ganglion). Note the abundant, agranular endoplasmic reticulum in the satellite cell of *Xenopus*. The *crossed arrow* points to a subsurface cisterna; **N** neuronal perikaryon; **sc** satellite cell. Transmission electron micrographs, (**a, b**) ×39,500; (**c**) ×23,800

diameter) as to distort the outline of the nucleus and even that of the whole cell (Rosenbluth 1962a; Arnold 1970).

In rabbit spinal ganglia, large lacunae (Fig. 1.14) have been occasionally observed beneath the outer surface of the satellite cell sheath. A continuity between the lacunar space and the perineuronal connective tissue space has not been traced so far. The largest lacuna observed measured 12 × 6 μm in section. Sometimes the lacuna appears partitioned by thin cytoplasmic septa devoid of organelles. Microvilli and sometimes also the shaft of a cilium may project into the lacunar space (Fig. 1.14). An electron transparent or a flocculent, moderately dense material fills the lacunar space. The significance of these cytoplasmic lacunae is unclear. So far they have been described only in the rabbit.

The plasma membrane of satellite cells, like that of neuroglial cells in the CNS, appears thinner and less dense than that of neurons (Figs. 1.14, 1.18 and 1.20). Assemblies of orthogonally arranged particles have been found in the freeze-fractured plasma membrane of satellite cells (Fig. 1.15). The center-to-center spacing of these particles is about 7 nm. Such assemblies have been found, for example, in the satellite cell plasma membrane of the chick (Pannese et al. 1977) and rat (Gotow et al. 1985) but not the frog (Matsumoto and Rosenbluth 1986). Similar specializations also are found in astrocytes, ependymal cells of higher vertebrates, and other cell types, whereas they have never been recognized within the plasma membranes of oligodendrocytes or neurons. The number per unit surface area of these assemblies is always lower in the plasma membrane of satellite cells than in that of astrocytes. These assemblies are composed of aquaporin-4 (Verbavatz et al. 1997; Rash et al. 1998). Aquaporins are channel-forming membrane proteins that are involved in water transport across the plasma membrane. Unlike aquaporin-4, other aquaporins do not usually form assemblies of orthogonally arranged particles.

1.5 Molecular Characteristics of Satellite Cells

Knowledge in this field is still fragmentary. A number of molecules found in satellite cells are listed in Table 1.1.

Under normal conditions, the perineuronal satellite cells of sensory ganglia express low levels of glial fibrillary acidic protein (see Table 1.1 for references),

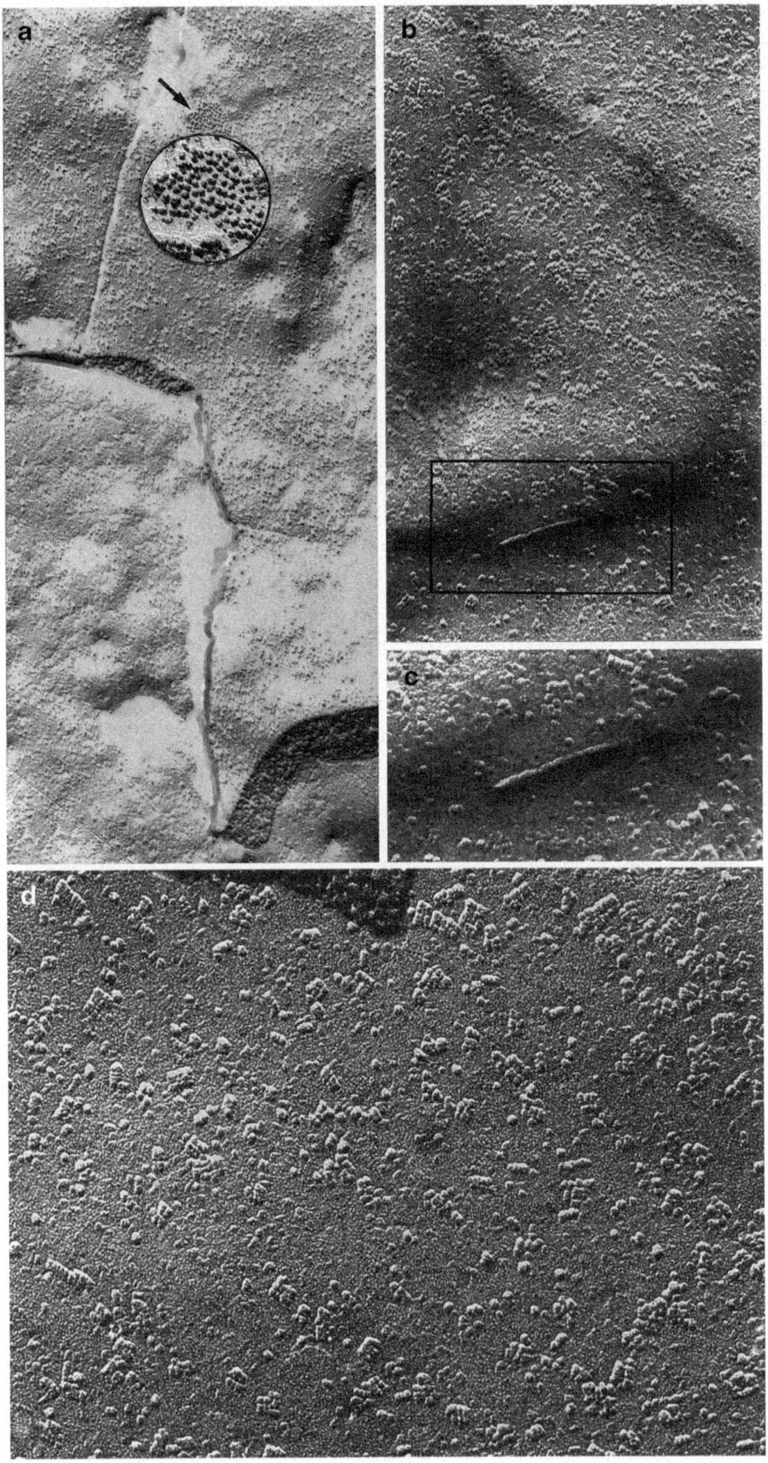
a
b
c
d

Fig. 1.15 Plasma membrane specializations of satellite cells (fowl spinal ganglion). A gap junction (**a**, *arrow*), a ridge, possibly a rudiment of a tight junction (**b**, **c**), and assemblies of orthogonally arranged particles (**d**) can be seen on the P face of the split plasma membrane of satellite cells. Freeze-fracture replicas (preparations by E. Reale, L. Luciano, and E. Pannese). (**a**) ×40,000; inset ×100,000; (**b**) ×60,000; (**c**, **d**) ×100,000

Table 1.1 A partial list of molecules identified in satellite cells of sensory ganglia

Molecule	Ganglion	Species	Reference
Glial fibrillary acidic protein	Spinal ganglia	Rat (adult)	Woodham et al. (1989), Fenzi et al. (2001), Jimenez-Andrade et al. (2006)
	Trigeminal ganglion	Rat (adult)	Stephenson and Byers (1995), Chudler et al. (1997), Ajima et al. (2001), Jimenez-Andrade et al. (2006)
	Nodose ganglion	Mouse (adult)	Feldman-Goriachnik et al. (2015)
S100 proteins	Spinal ganglia (in culture)	Quail (embryo)	Holton and Weston (1982)
	Spinal ganglia	Rat (adult)	Stefansson et al. (1982), Vega et al. (1989), Aoki et al. (1991), Sandelin et al. (2004)
	Spinal ganglia	Human[a]	Stefansson et al. (1982)
	Spiral ganglion	Rat (adult)	Aoki et al. (1991)
	Spiral ganglion	Human (adult)	Liu et al. (2014)
27-kD heat-shock protein	Spinal, trigeminal and nodose ganglia	Rat (adult)	Yamamoto et al. (2001)
Cyclic guanosine 3′-5′-monophosphate	Spinal ganglia	Rat (adult)	Morris et al. (1992), Aoki et al. (1993)
	Spinal ganglia	Rat (newborn)	Morris et al. (1992)
	Spinal and nodose ganglia (in culture)	Rat (adult)	Magnusson et al. (2000)
Cholecystokinin	Spinal ganglia	Bottlenose dolphin	Bombardi et al. (2010)
L-Arginine	Spinal and spiral ganglia	Rat (adult)	Aoki et al. (1991)
Annexin IV	Spinal ganglia	Rat (adult)	Naciff et al. (1996)
Nerve growth factor	Spinal ganglia	Rat (embryo)	Gill and Windebank (1998)
	Spinal ganglia (after axon injury)	Rat (adult)	Zhou et al. (1999)
Glial cell line-derived neurotrophic factor	Spinal ganglia (after axon injury)	Rat (adult)	Hammarberg et al. (1996)

(continued)

Table 1.1 (continued)

Molecule	Ganglion	Species	Reference
Neurotrophin-3	Spinal ganglia (after axon injury)	Rat (adult)	Zhou et al. (1999)
Brain-derived neurotrophic factor	Spinal, trigeminal and nodose ganglia	Rat (adult)	Wetmore and Olson (1995)
Transforming growth factors	Spinal ganglia	Rat (adult)	Stewart et al. (1995), Xian and Zhou (1999)
Fibroblast growth factors	Spinal ganglia	Rat (adult)	Grothe et al. (1997)
Glutamate dehydrogenase, glutamine synthetase, pyruvate carboxylase	Spinal ganglia	Rat[a]	Miller et al. (2002)
Heme oxygenase	Spinal and nodose ganglia	Rat (adult)	Magnusson et al. (2000)
Guanylate cyclase	Spinal and trigeminal ganglia	Guinea pig (adult)	Kummer et al. (1996)
Prostaglandin D synthase	Spinal ganglia	Domestic fowl (adult)	Vesin et al. (1995)
Cytidine monophosphatase, thiamine pyrophosphatase, nicotinamide adenine dinucleotide phosphatase	Spinal ganglia	Bullfrog	Bennett and Hemming (1989)
Ecto-nucleoside triphosphate diphosphohydrolase 2	Spinal ganglia	Rat[a]	Braun et al. (2004)
3-Phosphoglycerate dehydrogenase	Spinal ganglia	Rat (adult)	Yamashita et al. (2003), Kiya et al. (2011)
Tyrosine kinases	Spinal ganglia	Rat (adult)	Campana and Myers (2003)
Extracellular signal-regulated protein kinase	Spinal ganglia	Rat (adult)	Averill et al. (2001)
	Trigeminal ganglion	Rat (adult)	Freeman et al. (2008)
	Spinal ganglia (after axon injury)	Mouse (adult)	Rashid et al. (2004)
Mitogen-activated protein kinase	Spinal ganglia (in culture)	Mouse (adult)	Sjögreen et al. (2000)
Glutamate transporters	Spinal ganglia	Rat (adult)	Schon and Kelly (1974a, b), Berger and Hediger (2000), Carozzi et al. (2008)
	Spinal ganglia (in culture)	Rat (fetal)	Hösli and Hösli (1978)
	Nodose ganglion	Rat (adult)	Shoji et al. (2010)
γ-Aminobutyric acid transporter	Spinal ganglia	Rat (adult)	Schon and Kelly (1974a)
	Spinal ganglia (in culture)	Rat (fetal)	Hösli and Hösli (1978)
Peptide transporter 2	Spinal ganglia	Rat (adult)	Groneberg et al. (2001)

Table 1.1 (continued)

Molecule	Ganglion	Species	Reference
NKCC1 cotransporter	Spinal and trigeminal ganglia	Rat (adult)	Price et al. (2006)
Acetylcholine muscarinic receptors	Spinal ganglia	Domestic fowl (adult)	Bernardini et al. (1998)
	Spinal ganglia	Rat (adult)	Bernardini et al. (1999)
	Spinal ganglia	Rat (newborn)	Tata et al. (1999)
Purinergic receptors	Trigeminal ganglion	Mouse (adult)	Weick et al. (2003)
	Trigeminal ganglion	Rat (adult)	Vit et al. (2006)
	Spinal ganglia	Rat (adult)	Kobayashi et al. (2006)
	Spinal ganglia (in culture)	Rat[a]	Zhang et al. (2005)
Endothelin receptors	Spinal ganglia	Rat (newborn and adult)	Kar et al. (1991)
	Spinal ganglia	Rat[a], Rabbit[a], Monkey[a]	Pomonis et al. (2001)
Cytokine receptors	Spinal ganglia	Rat (adult)	Wetmore and Olson (1995), Zhou et al. (1996), Copray et al. (2001), Pannese and Procacci (2002), Campana and Myers (2003)
	Spinal ganglia (in culture)	Domestic fowl (embryo)	Zimmermann and Sutter (1983)
	Spinal ganglia (after axon injury)	Mouse (adult)	Ohtori et al. (2004)
	Trigeminal and nodose ganglia	Rat (adult)	Wetmore and Olson (1995)
Aspartate receptors	Spinal ganglia	Rat (adult)	Castillo et al. (2013)
Receptors to neuropeptides	Nodose ganglion	Human[a]	Burdyga et al. (2003)
	Spinal ganglia	Rat (adult)	Bär et al. (2004)
	Trigeminal ganglion	Rat (adult)	Li et al. (2008)
	Trigeminal ganglion (in culture)	Rat (newborn)	Vause and Durham (2010)
Bradykinin receptor	Spinal ganglia (after axon injury)	Mouse (adult)	Rashid et al. (2004)
γ-Aminobutyric acid receptors	Trigeminal ganglion	Rat[a]	Takeda et al. (2015)
Metabotropic glutamate receptors	Spinal ganglia	Rat (adult)	Carlton and Hargett (2007), Kung et al. (2013)
	Trigeminal ganglion	Rat (adult)	Kung et al. (2013)

(continued)

Table 1.1 (continued)

Molecule	Ganglion	Species	Reference
Epidermal growth factor receptor	Spinal ganglia	Human (fetal, young, and adult)	Werner et al. (1988)
	Spinal ganglia	Human (adult)	Huerta et al. (1996)
	Spinal ganglia	Rat (adult)	Xian and Zhou (1999)
Transcription factors	Spinal ganglia	Rat (adult)	Koike et al. (2014)
	Spinal ganglia	Rat (embryo)	Hagedorn et al. (2000)
Toll-like receptors	Spinal ganglia	Mouse (adult)	Kim et al. (2011)
	Spinal ganglia	Rat (adult)	Tse et al. (2014)
	Trigeminal ganglion	Human (adult)	Mitterreiter et al. (2017)
Major histocompatibility complex class II antigens	Spinal ganglia	Rat (adult)	Gehrmann et al. (1991)
	Spinal ganglia	Human (adult)	Esiri and Reading (1989), Graus et al. (1990)
	Trigeminal ganglion	Human (adult)	Esiri and Reading (1989)
Cell adhesion molecules	Spinal ganglia	Rat (adult)	Mirsky et al. (1986)
Dystroglycan complex	Spinal ganglia	Rat (adult)	Masaki et al. (2001)
Inwardly rectifying K^+ channels	Cochlear ganglion	Rat (adult)	Hibino et al. (1999)
	Spinal and trigeminal ganglia	Rat (adult)	Vit et al. (2006)
Acid-sensing ion channel 2	Spinal ganglia	Rat (adult)	Kawamata et al. (2006)

[a]Age not indicated in the original paper

while the satellite cells of autonomic ganglia express high levels of this protein (Elfvin et al. 1987; Alvarez et al. 1989). The glial fibrillary acidic protein of satellite cells has the same molecular weight as that of astrocytes, but their molecular identities differ slightly (Jessen et al. 1984). The expression of this protein increases markedly in a variety of conditions, e.g., following axon injury (Elfvin et al. 1987; Woodham et al. 1989; Stephenson and Byers 1995; Chudler et al. 1997; Vit et al. 2006; Takeda et al. 2007), compression of the ganglion (Zhang et al. 2009), infusion of paclitaxel (Jimenez-Andrade et al. 2006), or intraperitoneal administration of lipopolysaccharide (Feldman-Goriachnik et al. 2015). The expression of the following molecules also increases in satellite cells after axon injury: S100A4 protein (Sandelin et al. 2004), transforming growth factors (Xian and Zhou 1999), fibroblast growth factors (Grothe et al. 1997), heme oxygenases (Magnusson et al. 2000), receptors to neurotrophins (Zhou et al. 1996), tumor necrosis factor-α (Ohtori et al. 2004), cell adhesion molecules (Zhang et al. 2000), extracellular signal-regulated protein kinase (Rashid et al. 2004), interleukin 1β (Takeda et al. 2007), and major histocompatibility complex class II antigens (Gehrmann et al. 1991).

When intercellular relations are disrupted (e.g., in dissociated cell culture), molecular markers must be used to identify satellite cells. Among these are the S100

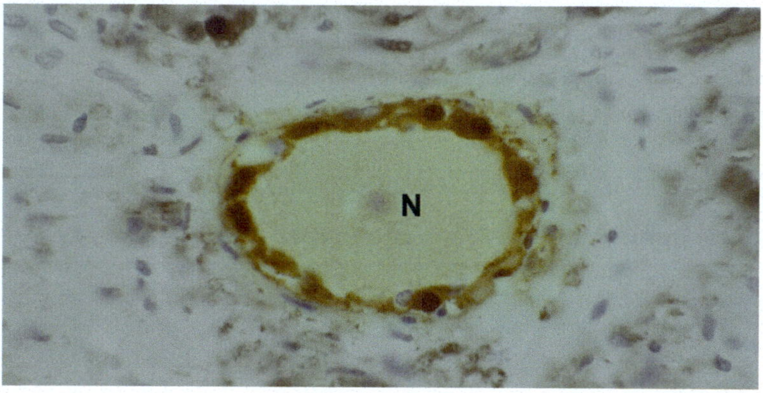

Fig. 1.16 Perineuronal satellite cells expressing S100 proteins (human spinal ganglion). **N**, nerve cell body. Light micrograph ×450 (Courtesy of F. Scaravilli)

proteins (Fig. 1.16) which, however, are also expressed by Schwann cells. More selective satellite cell markers are the glutamate-aspartate transporter and glutamine synthetase.

Guanylate cyclase has been located within satellite cells (Kummer et al. 1996). This enzyme is activated by nitric oxide released by neurons and catalyzes the synthesis of cyclic guanosine 3',5'-monophosphate. Following axon lesion, the synthesis of the latter enzyme by satellite cells increases (Shi et al. 1998). Satellite cells contain arginine, the precursor of nitric oxide. It has been suggested that satellite cells supply arginine to neurons for nitric oxide production (Aoki et al. 1991). Other molecules that satellite cells synthesize are cytokine tumor necrosis factor-α (Shimeld et al. 1997; Ohtori et al. 2004) and prostaglandins (Vesin et al. 1995).

Table 1.1 shows that the satellite cells of sensory ganglia are endowed with receptors for a variety of molecules. These cells are thus able to transport these molecules into the cell body. This ability is important for controlling the composition of the perineuronal microenvironment (see Sect. 1.16.2).

1.6 Relationships Between Satellite Cells

In certain areas, adjacent satellite cells contact simply, while in other areas within the same sheath, their lamellar expansions overlap and interdigitate in a complex fashion (Fig. 1.6). Where numerous lamellar expansions are overlapped, more than one may belong to the same cell, and thus the overlapping satellite cells are sometimes less numerous than the lamellar expansions observed. Clefts, from 15 to 20 nm in width, intervene between the apposed plasma membranes of adjacent satellite cells (Fig. 1.6). These clefts open into the light space under the basal lamina at one end and into the narrow interval between the neuron and the sheath at the other. This potential pathway between the interstitial connective tissue of the ganglion and the neuronal surface usually follows a long and tortuous course even in the thin regions of the sheath, a condition that probably has some functional significance (see Sect. 1.16.1).

Satellite cells are linked to other satellite cells within the same perineuronal sheath by adherent, gap, and rudimentary tight junctions (Pannese 1974; Pannese et al. 1977, 1978; Sakuma et al. 2001; Liu et al. 2014). Adherent junctions are small, button-like structures. Adherent junctions have been described also between adjacent Schwann cells (Eames and Gamble 1970). Macular gap junctions (Figs. 1.15 and 1.26) between satellite cells have been revealed with the freeze-fracture technique (Pannese et al. 1977, 1978). Gap junctions between satellite cells have also been found in autonomic ganglia (Elfvin and Forsman 1978). Strands, which appear as ridges on the P face (Figs. 1.15 and 1.26) and as grooves on the E face (Pannese et al. 1978), have been observed in the freeze-fractured plasma membrane of satellite cells. Since these strands show the same morphological characteristics as those composing tight junctions in other tissues, they could be focal tight junctions (maculae occludentes) linking adjacent satellite cells. Strands have been revealed also in the replicas of the split plasma membrane of satellite cells in autonomic ganglia (Elfvin and Forsman 1978). Also such strands have been interpreted as rudiments of tight junctions.

All the above types of junction probably contribute to the adhesion between satellite cells. Furthermore, gap junctions have other important roles. The internal diameter (about 1.2 nm) of the gap junction channels not only allows the intercellular passage of ions but also permits the transfer of molecules with molecular weight up to about 1 kD (e.g., signaling molecules, amino acids, and glucose). Dye-coupling studies have shown that satellite cells enveloping a given nerve cell body are functionally coupled (Hanani et al. 2002; Pannese et al. 2003; Huang et al. 2005, 2006). Information on the connexins that form these junctions is scarce. Connexin43 has been identified in the perineuronal satellite cells of rat trigeminal ganglion (Vit et al. 2006), of mouse spinal ganglia (Procacci et al. 2008), and of guinea pig and human spiral ganglion (Liu et al. 2014); connexin36 has been identified in the perineuronal satellite cells of rat spinal ganglia (Pérez Armendariz et al. 2018). Reflexive adherent and reflexive gap junctions, i.e., junctions linking different portions of the same satellite cell, have also been described.

1.7 Perikaryal Myelin Sheaths

Myelin sheaths enveloping nerve cell bodies in sensory ganglia were described in the nineteenth century (Wagner 1846; Bidder 1847; Leydig 1851; Schramm 1864; Courvoisier 1868; Schultze 1871 and other later authors). Under polarized light, these perikaryal sheaths present the same characteristics shown by the myelin surrounding the axon. Myelinated nerve cell bodies occur frequently in the eight cranial nerve ganglia in all classes of vertebrates and occasionally in other sensory ganglia of bony fishes (for a review, see Scharf 1958). Most of the myelinated nerve cell bodies are bipolar, some pseudounipolar. In the autonomic nervous system, myelinated nerve cell bodies occur in the ciliary ganglion of birds.

Electron microscope studies by Rosenbluth and Palay (1961), Rosenbluth (1962b), and subsequently other authors showed that the perikaryal myelin may display either the compact structure exhibited by the myelin surrounding the axons or the characteristics of loose myelin. In the latter case, the perikaryal sheath is built of a varying number of lamellae (from 10 to 100 nm in thickness), each consisting of a layer of satellite cell cytoplasm bounded by its plasma membrane and separated from the nearby lamellae by a narrow space (Fig. 1.17). Compact myelin has a highly regular pattern of organization,

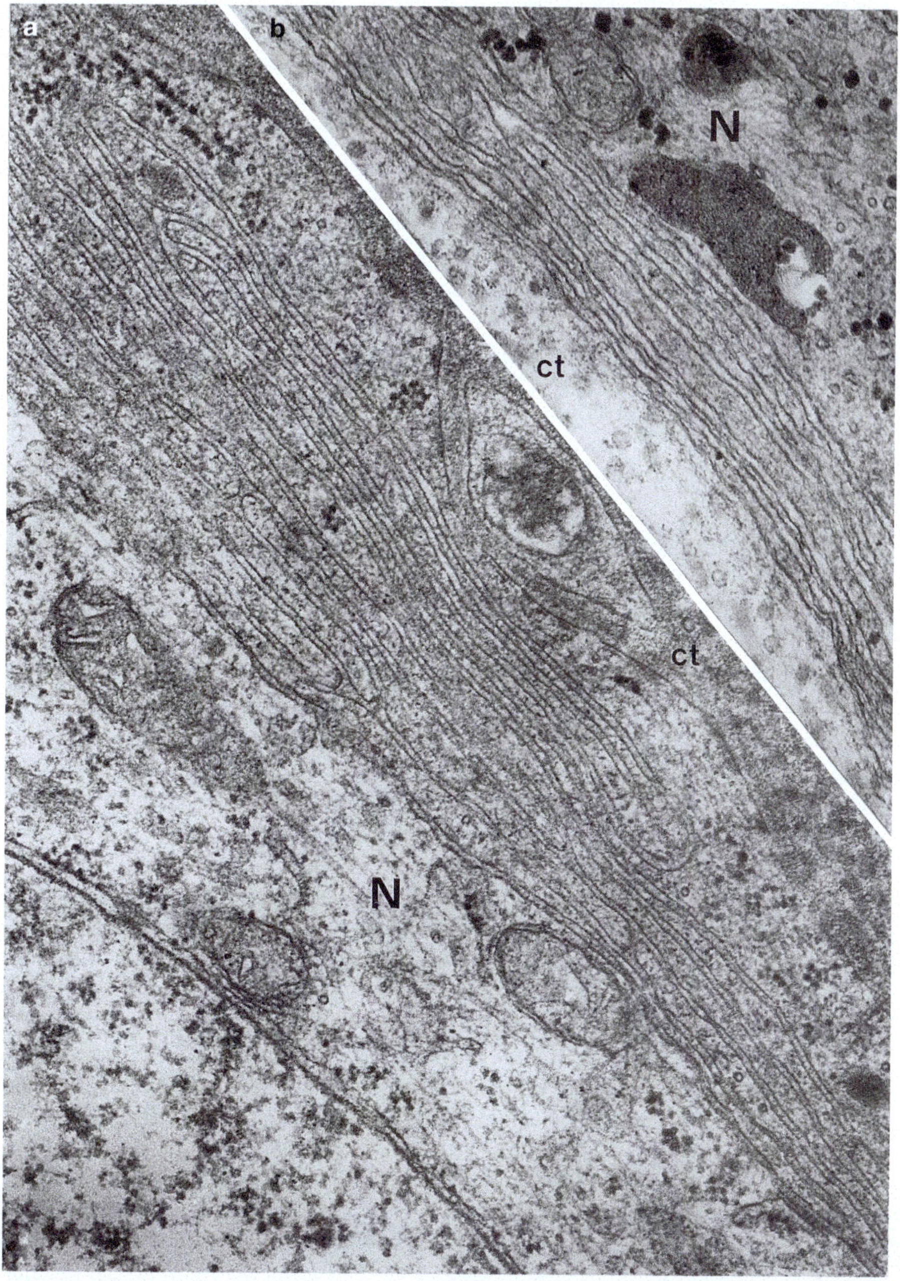

Fig. 1.17 Perikaryal myelin sheaths in the spinal ganglia of a cat (**a**) and a bullfrog (**b**). **ct** connective tissue; **N** neuronal perikaryon. Transmission electron micrographs, ×48,000

whereas in loose myelin the number and thickness of the lamellae and the width of the interval between the lamellae in a given sheath may vary from one region to another. The main characteristics of perikaryal loose myelin are similar to those of the sheaths enveloping certain axons of invertebrates and the developing myelinated nerve fibers of vertebrates. Perikaryal loose myelin differs from common multilayered perikaryal sheaths only in the thickness of the cytoplasmic layers and the location of the cell nuclei; in loose myelin, the cytoplasmic layers are thinner than in common multilayered perikaryal sheaths, and the satellite cell nuclei always lie in the outermost layer of the sheath. Between compact and loose myelin, intermediate types are found. Both compact and loose myelin may coexist in the same perikaryal sheath.

Thus, a wide range of perikaryal sheaths can be found in sensory ganglia, namely, a simple layer of satellite cells, multiple layers of satellite cells, loose myelin, loose and compact myelin coexisting in the same sheath, and, finally, compact myelin. Therefore, the perikaryal myelin sheaths are not basically different from the satellite cell envelopes usually found around the nerve cell bodies but rather represent variants of them. Nerve cell bodies surrounded by a myelin sheath also occur in the CNS (for further details, see Pannese 2015).

1.8 Neuron-Satellite Cell Boundary

The neuron-satellite cell boundary is usually complicated by the presence of many slender projections from both the neuron and the satellite cells (Hess 1955; Palay 1957; Wyburn 1958; and later authors). In single sections, the neuronal projections may sometimes appear continuous with the neuronal perikaryon (Figs. 1.5, 1.10, 1.11, 1.12, and 1.18); more frequently, they arise from the neuron at another level, thus appearing as discrete entities completely surrounded by satellite cells (Figs. 1.5, 1.6, 1.18, and 1.21), or located along the interval between the neuron and its satellite cell sheath. Even in these cases, however, they can be identified because their cytoplasmic matrix is less dense and their plasma membrane thicker and denser than their counterparts in the adjacent satellite cells. In sections grazing the neuronal surface (Fig. 1.18) and in neurons deprived of their enveloping sheaths (Fig. 1.19), one can realize how numerous such projections are. The outgrowth of these projections is influenced by the satellite cells investing the nerve cell body (see Sect. 1.16.4). Neuronal projections, similar to those above, have been described also in autonomic ganglia (Pick 1963; Forssmann 1964).

Satellite cell projections, which occur less frequently than neuronal ones, invaginate the neuronal profile to variable depths (Fig. 1.20); these projections possibly correspond to the trophospongium described by Holmgren (1901, 1902). The plasma membrane of the sensory neuron may exhibit deep invaginations which do not contain projections from satellite cells (Fig. 1.20). These invaginations appear similar to those observed in large neurons of invertebrates (see, among others, Rosenbluth 1963; Coggeshall 1967).

Both smooth-surfaced and coated pinocytotic vesicles can be frequently found along both the neuronal plasma membrane (Fig. 1.18) and the satellite cell plasma

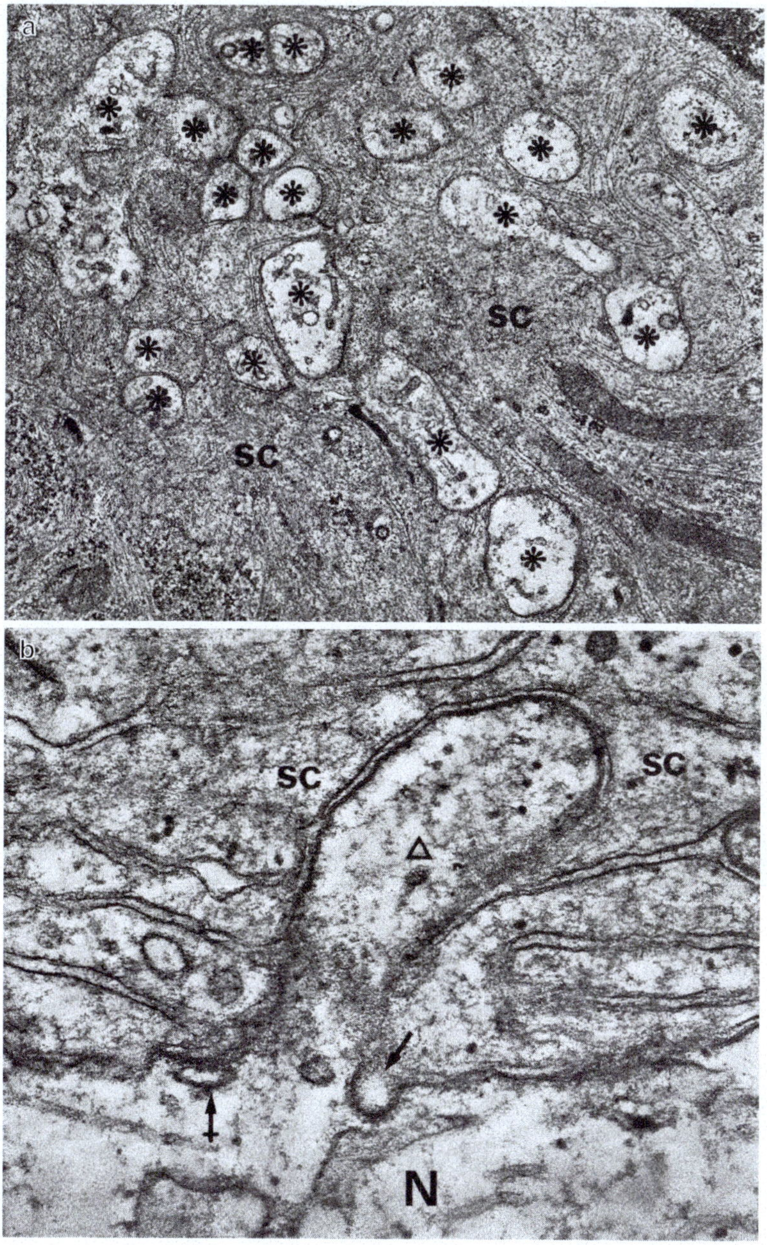

Fig. 1.18 Neuronal projections. (**a**) Section grazing the surface of a nerve cell body (cat spinal ganglion). *indicates cross sections of neuronal projections; **sc** satellite cell. (**b**) Section almost perpendicular to the surface of a nerve cell body (fowl spinal ganglion). *Open triangle* indicates a projection which arises from the neuronal perikaryon (**N**) and invaginates the inner contour of the satellite cell sheath (**sc**). *Arrow* points to a coated pinocytotic vesicle; *crossed arrow* points to a subsurface cisterna. Note that the plasma membrane of the satellite cell is thinner and less dense than that of the neuron. Transmission electron micrographs, (**a**) ×21,000; (**b**) ×68,000

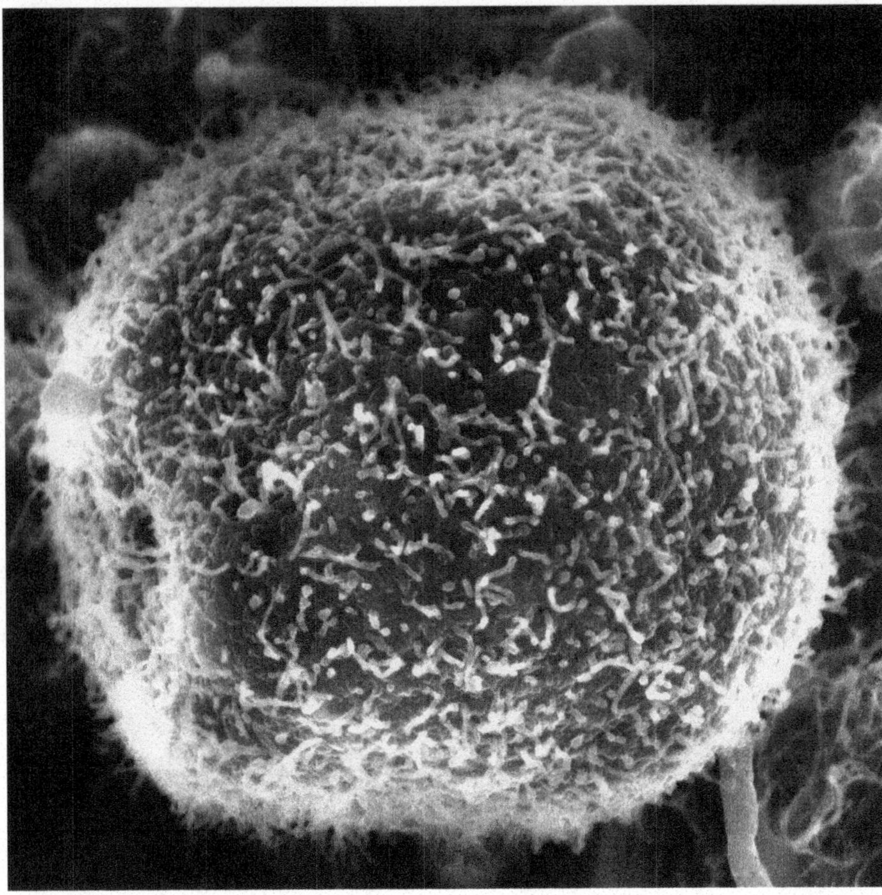

Fig. 1.19 Neuronal projections. The perikaryal surface of a sensory neuron is exposed after removal of all enveloping sheaths (rabbit spinal ganglion). The surface is rendered irregular by the presence of numerous projections. Scanning electron micrograph, ×3800

membrane facing the neuron (Fig. 1.20). Subsurface cisternae closely applied to the plasma membrane facing the satellite cell sheath (Figs. 1.12, 1.13, 1.14, 1.18, 1.20, and 1.21) have been described in the neurons of sensory ganglia. In satellite cells, subsurface cisternae can be rarely found apposed to the plasma membrane facing the neuron. Confronting subsurface cisternae can very rarely be seen in a neuron and in a satellite cell.

The interval between the satellite cell sheath and the enclosed neuron (Figs. 1.6, 1.12, 1.14, 1.18, 1.20, and 1.21) has a fairly uniform width (about 20 nm), except where numerous projections arise from the neuron; here it appears uneven and enlarged. Button-like, adherent junctions have occasionally been observed at the neuron-satellite cell boundary in spinal (Pannese 1969; Tennyson 1970), spiral (Adamo and Daigneault 1972), petrosal (Stensaas and Fidone 1977), and autonomic (Dixon 1966; Unsicker

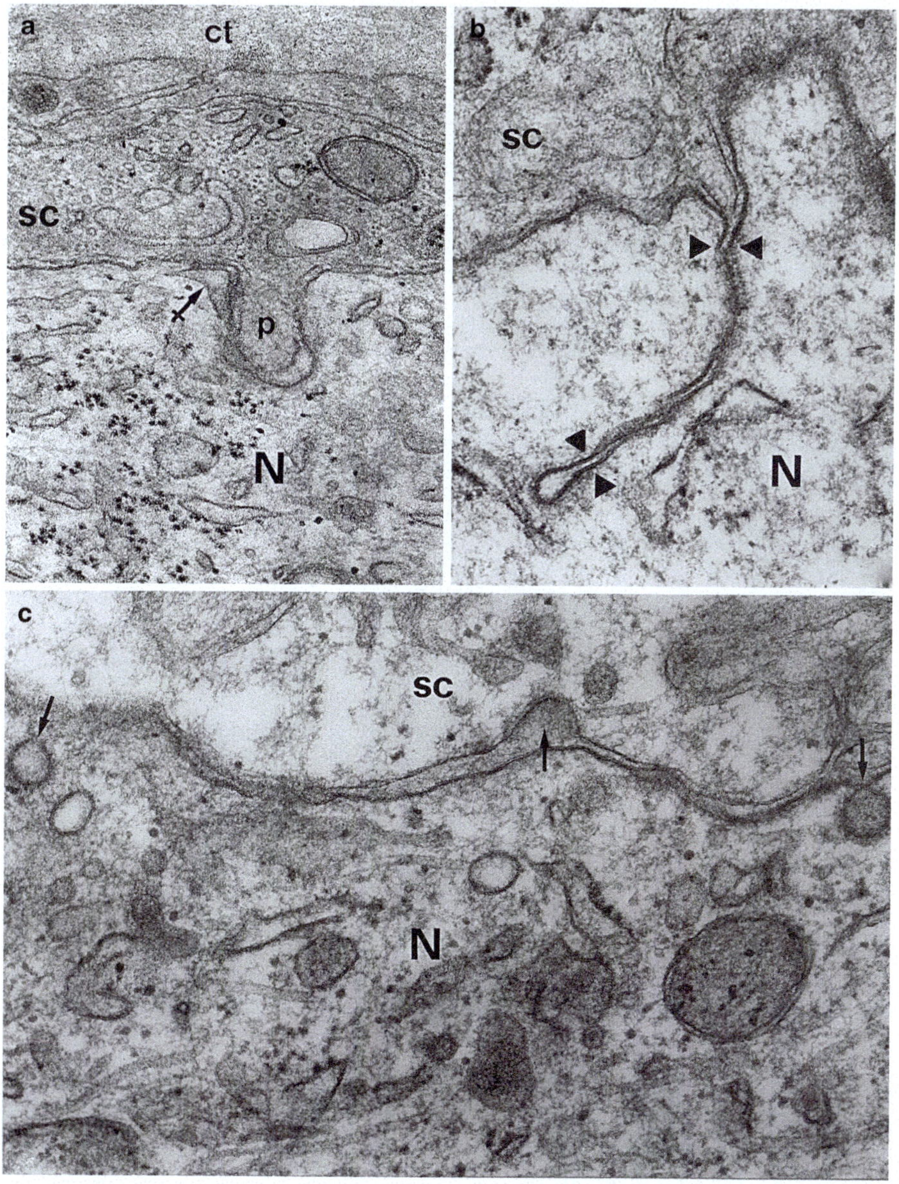

Fig. 1.20 Neuron-satellite cell boundary (spinal ganglia of rabbit (**a**) and fowl (**b**, **c**)). (**a**) A satellite cell projection (**p**) invaginates the neuronal profile. Cross sections of intermediate filaments and microtubules can be seen in satellite cells (**sc**); **ct** interstitial connective tissue; **N** neuronal perikaryon. *Crossed arrow* points to a subsurface cisterna. (**b**) Deep invagination (between *filled triangles*) of the neuronal plasma membrane which does not contain a satellite cell (**sc**) projection; **N** Neuronal perikaryon. (**c**) Coated pinocytotic vesicles (*arrows*) along both the neuronal plasma membrane and the satellite cell (**sc**) plasma membrane facing the neuronal perikaryon (**N**). Note that the plasma membrane of the satellite cell is thinner and less dense than that of the neuron. Transmission electron micrographs, (**a–c**) ×59,000

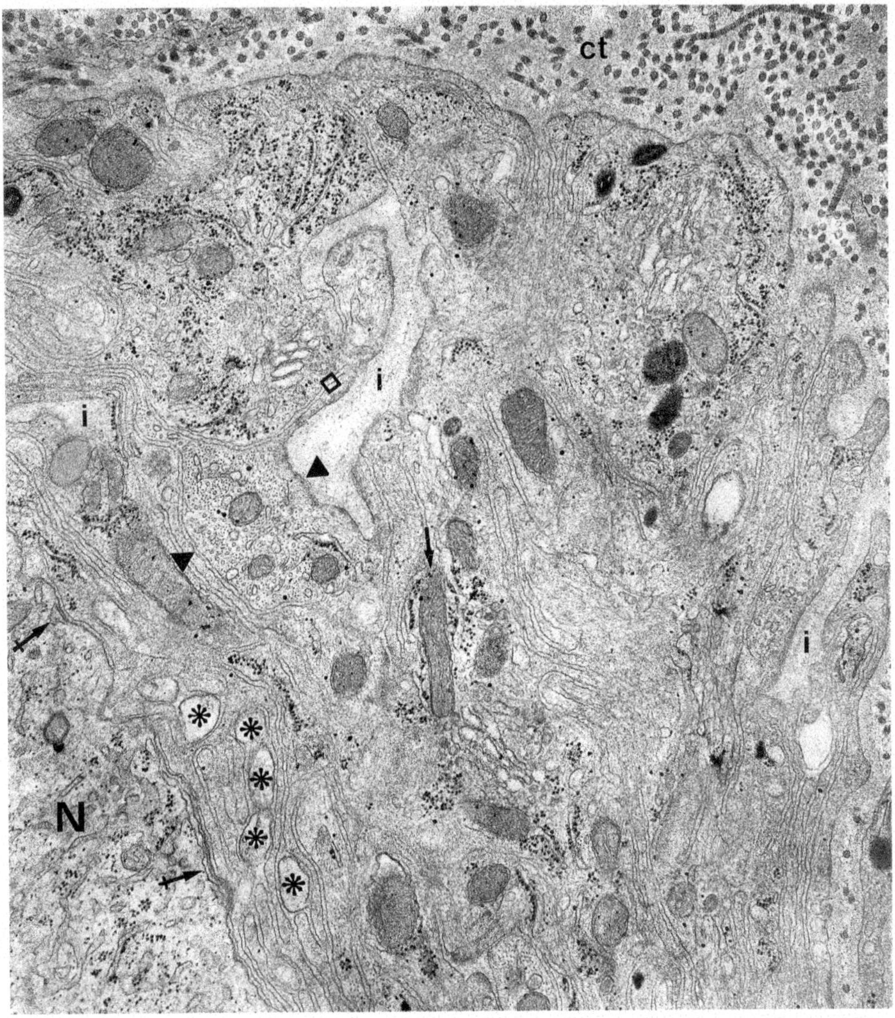

Fig. 1.21 Deep invaginations (**i**) of the outer contour of the satellite cell sheath (rabbit spinal ganglion). In this section, the invagination on the right communicates with the interstitial connective tissue (**ct**) of the ganglion. Cisternae of the granular endoplasmic reticulum, several sections of the Golgi apparatus, mitochondria, and dense bodies are evident in satellite cells; many cross-sectioned intermediate filaments and microtubules can be seen between ► and ◄. *Arrow* points to a mitochondrion with longitudinally arranged cristae. *Crossed arrows* point to subsurface cisternae; * indicates cross sections of projections which arise from the neuronal perikaryon (**N**) at other levels. Transmission electron micrograph, ×23,000

1967) ganglia. However, as far as I know, tight or gap junctions have never been described at this interface. Notwithstanding this, some authors take it for granted that gap junctions are present between satellite cells and sensory ganglion neurons.

In sensory ganglia, Ehrlich (1886), Ramón y Cajal (1890), and Dogiel (1896, 1897) described a plexus of unmyelinated fibers surrounding some nerve cell bodies. Some authors regarded these fibers as sympathetic fibers of extrinsic origin (coming from outside the ganglion), while others regarded them as unmyelinated branches of myelinated fibers from neurons situated within the ganglion (Dogiel type II cells). Although the presence of a perineuronal plexus of unmyelinated fibers in sensory ganglia was questioned by Van Gehuchten (1892) and Retzius (1894), the existence of this plexus has been repeatedly confirmed over the last 20 years. The perineuronal unmyelinated fibers are currently regarded as sympathetic fibers of extrinsic origin (for more details, see Matsuda et al. 2005). In normal conditions, only a small proportion of ganglionic sensory neurons is surrounded by sympathetic fibers. Sometimes these fibers are enclosed within the satellite cell sheath that envelopes the sensory nerve cell body, and very rarely these fibers are directly apposed to the plasma membrane of the sensory nerve cell body but never make synaptic contact with it. Since the number of sympathetic fibers surrounding ganglion sensory neurons increases notably after peripheral nerve injury, it has been suggested that this sympathetic innervation of ganglion sensory neurons may contribute to the generation and maintenance of some forms of neuropathic pain (see Sect. 1.18.4).

1.9 Boundary Between the Satellite Cell Sheath and the Interstitial Connective Tissue

The outer surface of the satellite cell sheath facing the interstitial connective tissue of the ganglion is covered by a basal lamina (Figs. 1.6, 1.11, 1.12, and 1.14), which is continuous from one satellite cell to the next, i.e., bridges the 15–20 nm gaps between adjacent satellite cells (Fig. 1.6). Notwithstanding the presence of evaginations and invaginations, the outer surface of the sheath is generally smoother than that facing the neuron. The evaginations of satellite cells appear as lamellar or fingerlike cytoplasmic projections which extend into the connective tissue and are always surrounded by the basal lamina. In single sections, the profiles of these evaginations may sometimes be seen as discrete entities in the connective tissue since they arise from the satellite cell sheath at another level; in these cases, they may be distinguished from cytoplasmic processes of the connective tissue cells as the latter are devoid of a basal lamina (Fig. 1.11).

Invaginations, which again always appear lined by a basal lamina, enter the satellite cell sheath to variable depths (Fig. 1.21). Such invaginations are generally deeper, more numerous, and more ramified around the initial segment of the axon than around the neuronal perikaryon; in the former location, they sometimes form a labyrinthine system which may nearly reach the axonal surface. In some amphibians and reptiles, the satellite cell sheath enveloping the initial segment of the axon is divided into two layers (Fig. 1.22); the intervening cleft, which may contain collagen fibrils, communicates with the interstitial connective tissue.

Dense plaques of amorphous material may be found apposed to the cytoplasmic surface of the satellite cell plasma membrane facing the basal lamina (Fig. 1.21). Where a dense plaque occurs, the plasma membrane usually protrudes from the remainder of the cell surface. No thickening or increased density is shown by the basal lamina facing the intracellular dense plaque. In the cytoplasmic matrix near

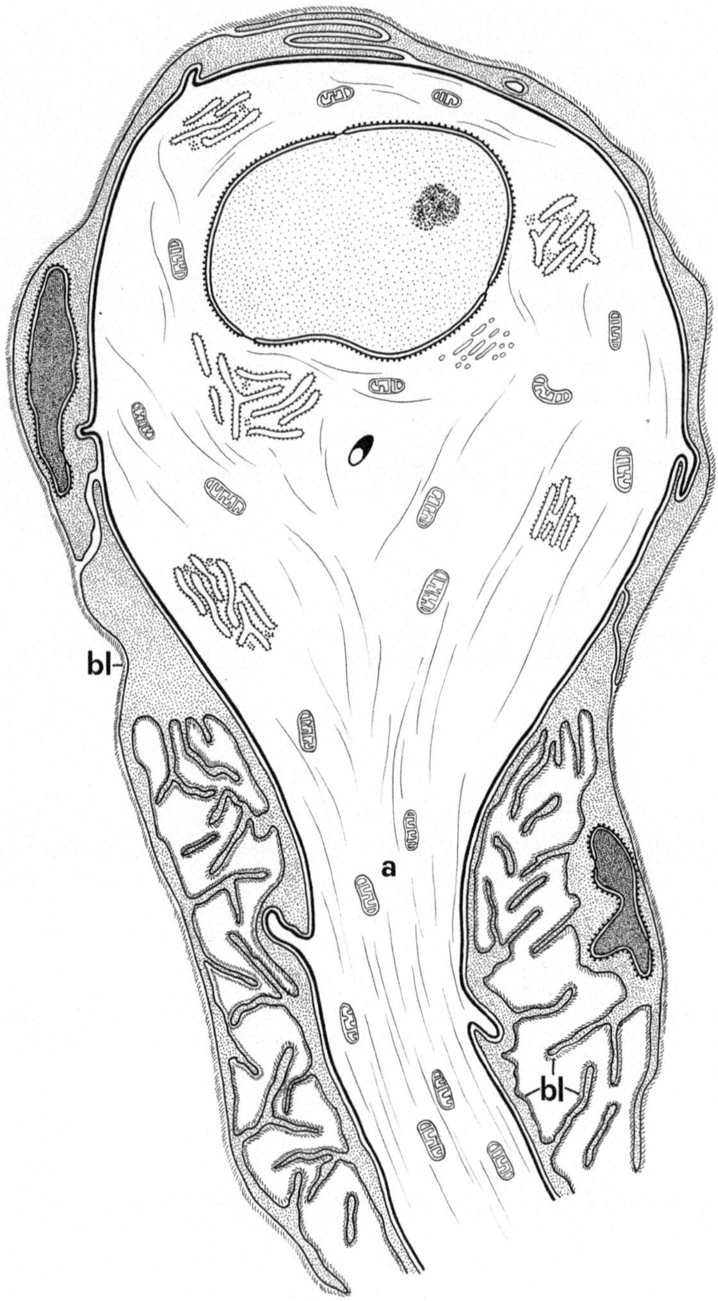

Fig. 1.22 Diagram showing a neuron of a bullfrog spinal ganglion with its satellite cell sheath. At the level of the initial segment of the axon (**a**), the satellite cell sheath (stippled) appears split into two layers; the intervening cleft communicates with the interstitial connective tissue of the ganglion; lamellar projections, rising from the confronting surfaces of the two layers, jut into this intervening space; **bl** basal lamina

the dense plaque, numerous filaments are usually found. These structures resemble those that Krawczyk and Wilgram (1973) described in the intermediate stage of hemidesmosome formation. The molecular constituents of these structures do not appear to have been characterized, and the function of these structures therefore remains unclear.

Blood capillaries are numerous in the interstitial connective tissue of sensory ganglia (Fig. 1.2). Most of these capillaries are of the continuous, unfenestrated variety but fenestrated capillaries have also been described (e.g., see Olsson 1971; Anzil et al. 1976; Jacobs et al. 1976; Arvidson 1979). Some capillaries lie in close proximity to perineuronal satellite cells (Figs. 1.2, 1.3, 1.4, and 1.29). Unlike the endothelial cells of CNS capillaries, the endothelial cells of the capillaries within sensory ganglia have no tight junctions (see also Sect. 1.16.1).

1.10 Quantitative Relationships Between Nerve and Satellite Cells

Sensory ganglia are a particularly suitable material for studying the quantitative relationships between nerve and neuroglial tissues because both neuronal bodies and satellite cell sheaths have a rather regular shape. Moreover, unlike the situation in other regions of the nervous system, some connective tissue sharply separates the satellite cell sheath belonging to a neuron from the sheaths belonging to other neurons (Figs. 1.3 and 1.29), so that each nerve cell body and its related satellite cell sheath usually form a discrete structural unit (Pannese 1981). In sensory ganglia it is therefore possible to determine the quantitative relationships between individual nerve cells and their satellite cells.

Light microscope studies showed that in spinal ganglia the number of satellite cells enveloping each nerve cell body is proportional to the volume of the latter (Pannese 1960, 1964). Studies carried out with the electron microscope have revealed that in spinal ganglia the volume of the satellite cell sheath is directly proportional to both the volume and surface area of the nerve cell body (Pannese et al. 1972, 1975). The surface of each nerve cell body is highly irregular (Fig. 1.19), its area thus greatly exceeding that of a sphere of equal volume; this explains why the volume of the satellite cell sheath can be directly proportional to both the volume and the surface area of the nerve cell body. The number of satellite cells associated with each neuron correlates highly with the neuronal volume also in autonomic ganglia (Gabella et al. 1988; Pomeroy et al. 1996).

As shown by the studies quoted above, a quantitative balance exists between nerve cell bodies and their associated satellite cell sheaths. The factors which maintain this balance are a matter of hypothesis. Since the experimentally induced enlargement of the nerve cell body is followed by an increase in volume of the satellite cell sheath (Pannese 1963, 1964; Humbertson et al. 1969), it is reasonable to suppose that the size and/or surface area of the nerve cell body influence and control the volume of the related satellite cell sheath. By comparing the quantitative results obtained from mammals and reptiles, it appears that, for a given size or surface area of the nerve cell body, the volume of the satellite cell sheath is greater in mammals than in reptiles (Pannese et al. 1972, 1975). It seems, therefore, that in addition to the size and surface area of the nerve cell body, other factors, for example, the metabolic rate of the ganglionic neuron, may influence the volume of the satellite cell sheath.

As is the case for satellite cell sheaths and nerve cell bodies in spinal ganglia, in both myelinated and unmyelinated nerve fibers of spinal roots, the size of the Schwann cell sheath is directly proportional to the size of the related axon or axons (Pannese et al. 1987, 1988).

1.11 Origin and Development of Satellite Cells

Satellite cells of sensory ganglia share a common ectodermal origin with ganglion nerve cells (Dohrn 1891; Koelliker 1905; Streeter 1905, 1912; Kohn 1907; Lenhossék 1907; Levi 1907, 1908; Held 1909). The satellite cells of spinal ganglia originate from neural crest (Harrison 1904; Detwiler 1937; Jones 1939 and other authors) perhaps with a small contribution from neuroepithelial cells which migrate from the spinal cord after the emigration of crest cells is complete (Sharma et al. 1995). Data on the origin of the satellite cells of the cranial sensory ganglia are conflicting. For example, the work carried out by Narayanan and Narayanan (1980) and by D'Amico-Martel and Noden (1983) suggests that the satellite cells of chick cranial sensory ganglia derive exclusively from neural crest. By contrast, the data presented by Hamburger (1961) suggest that the satellite cells of the sensory ganglion of the trigeminal nerve of the chick derive from both the neural crest and from the cephalic placodes. Furthermore, the work by Yntema (1937, 1943) indicates that at least some of the satellite cells of the cranial sensory ganglia of *Amblystoma punctatum* have a placodal origin. The molecular mechanisms that direct specification of neural crest cells into satellite cells have been summarized by Jacob (2015).

As regards the time of appearance of satellite cells, Levi (1908), who studied ganglion development in many species, came to the conclusion that neuronal differentiation in sensory ganglia precedes satellite cell differentiation. Electron microscope observations (Pannese 1969) support this conclusion and show that satellite cells can be recognized by electron microscopy at earlier stages than by light microscopy, e.g., at the fifth instead of the seventh (Brizzee 1949) or sixth (Yates 1961) incubation day in chick embryo spinal ganglia. Schwann cells differentiate somewhat later than satellite cells (Woodhoo et al. 2004).

At an early developmental stage of the sensory ganglion, undifferentiated cells are scattered singly or in small clusters among the immature neurons. Under the electron microscope, undifferentiated cells appear rounded or polyhedral with an ovoid or spherical nucleus encircled by a cytoplasmic rim (Fig. 1.23). The chromatin appears rather condensed and the nucleolus often shows a rather compact texture. Clusters of free ribosomes, small and sparse cisternae of the granular endoplasmic reticulum, a poorly developed Golgi complex, a few small mitochondria, and individual microtubules but

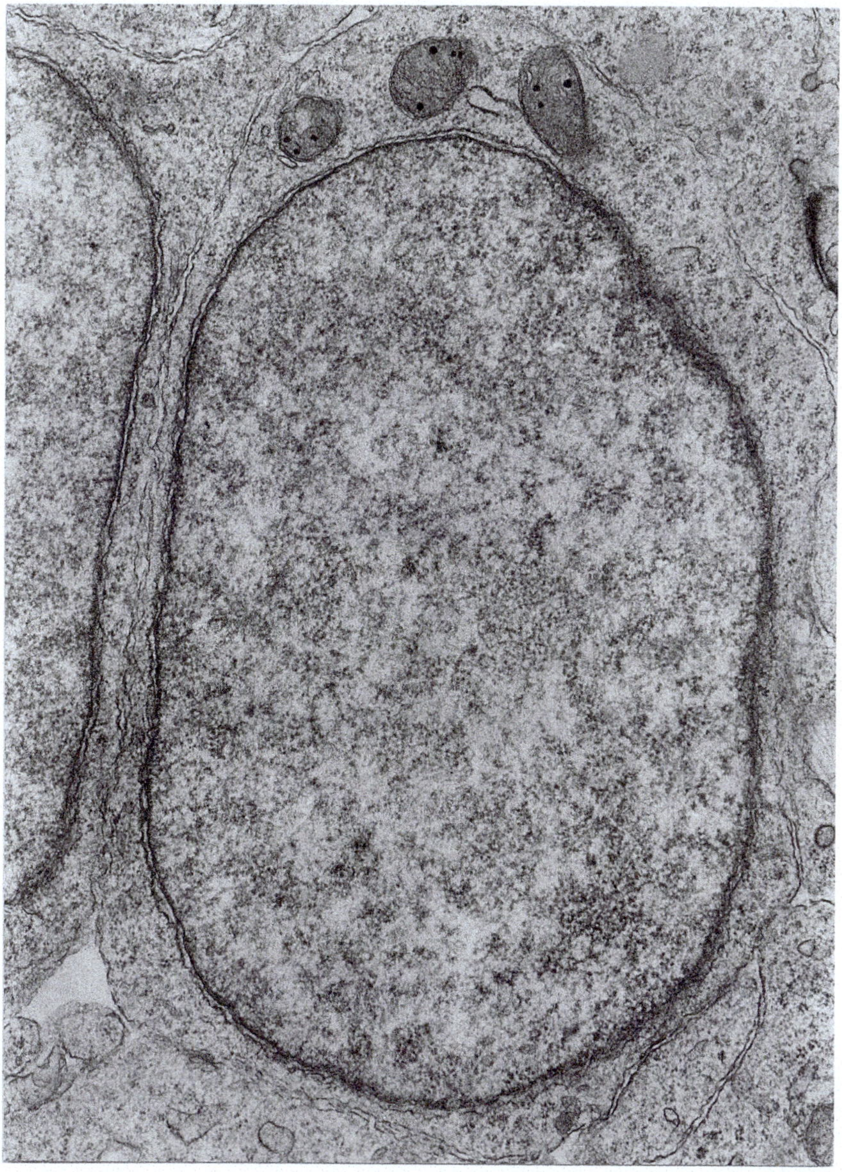

Fig. 1.23 Undifferentiated cell (chick embryo spinal ganglion). Transmission electron micrograph, ×24,000

not intermediate filaments are present in the cytoplasmic rim. Rarely a cilium is found which projects into a channel formed by a deep invagination of the plasma membrane. Adherent and gap junctions sometimes link the undifferentiated to adjacent cells. Undifferentiated cells can often be observed in mitosis (Fig. 1.24). The dividing undifferentiated cells retain the junctions with adjacent cells (Pannese 1968b; Gaik and

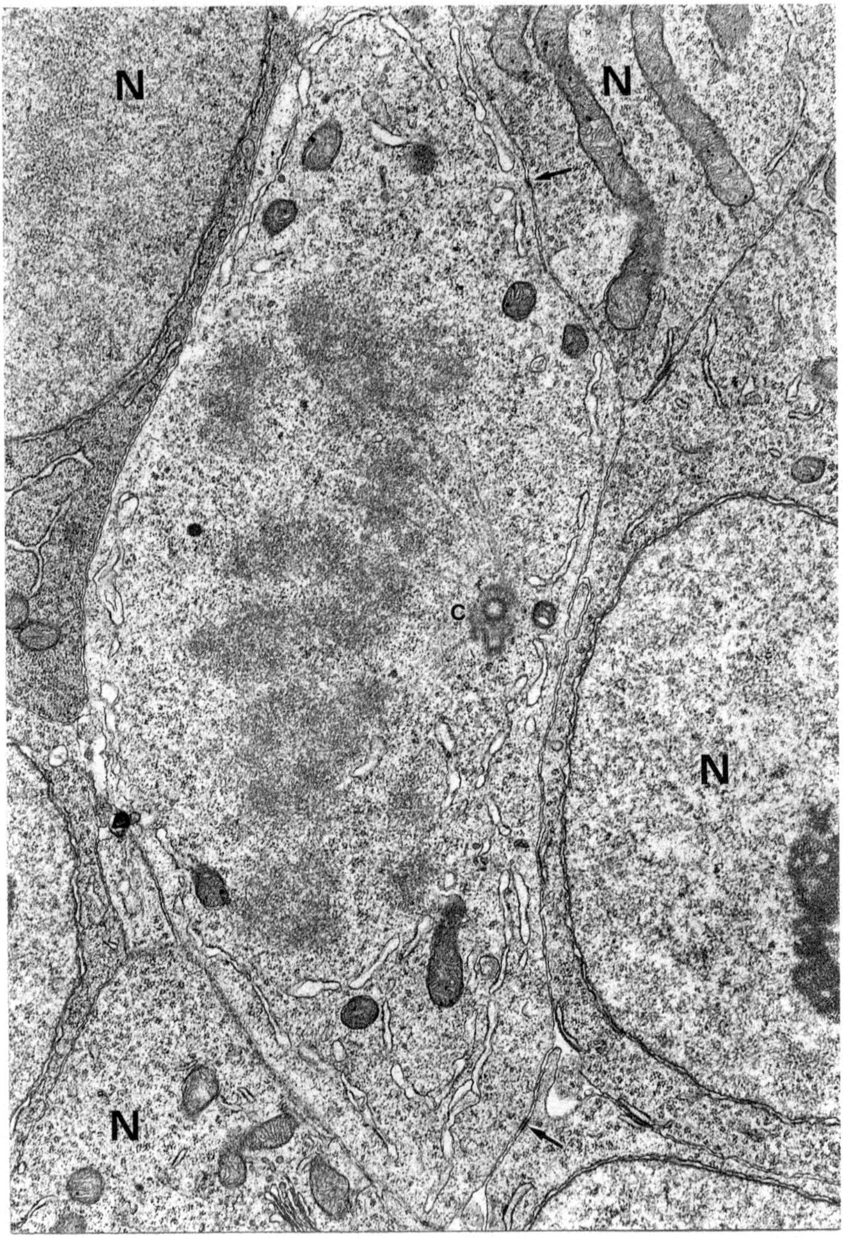

Fig. 1.24 Mitotic undifferentiated cell located in the center of a group of immature neurons (**N**) in a chick embryo spinal ganglion. *Arrows* point to adhering junctions, which link the mitotic cell to adjacent cells; **c** centrioles. Transmission electron micrograph, ×15,300

Farbman 1973), as also observed in neural tube by Hinds and Ruffett (1971). Findings suggest that signals arising in adjacent neurons promote differentiation of undifferentiated cells into satellite cells (Pruginin-Bluger et al. 1997).

On the basis of studies carried out mainly on spinal ganglia (Pannese 1969; Tennyson 1970), the following developmental stages of satellite cells can be traced. When satellite cells are first detected in the ganglion, they appear star shaped in section and consist of a nucleated body with several attenuated cytoplasmic expansions (Fig. 1.25). The body of each satellite cell is located in the center of a group of immature neurons and its cytoplasmic expansions radiate out intervening between the nerve cells. Thus, at this stage, each satellite cell is related to several nerve cells (Pannese 1969; Krajčí 1973); the nerve cells greatly outnumber satellite cells, and therefore only some come into direct contact with satellite cells. The majority of nerve cells remain in mutual contact. The satellite cell nucleus appears nearly triangular or more irregularly shaped in section. The nucleoplasm is somewhat denser than its counterpart in the nerve cell; the chromatin occurs in small clumps scattered throughout the nucleoplasm and in a thin peripheral accumulation. Clusters of free ribosomes, one or two short profiles of the granular endoplasmic reticulum, a few mitochondria, a Golgi complex, and some microtubules (20–30 nm in diameter) are usually seen in every section of the satellite cell body (Fig. 1.25); lysosomes, multivesicular bodies, dense bodies, centrioles, and lipid droplets are occasionally observed. Pinocytotic vesicles, mostly of the coated type, are often present along the plasma membrane. The plasma membrane of the thinnest expansions of the satellite cell may appear wrinkled. If this aspect is not artifactual but it reflects a dynamic condition preexisting in the living state, the mentioned expansions might be looked upon as undulating membranes. The satellite cell expansions can intervene between the immature neurons and separate them even where the latter are connected by adherent or gap junctions (Pannese 1968b, 1969; Pannese et al. 1977). It is unknown, however, whether the separation of adjoining nerve cells depends on the simple mechanical activity of satellite cells or whether these cells release substances that block the action of the adhesion molecules at these sites. At this stage, the ganglion still has an epithelium-like structure, its cells being separated everywhere by intercellular clefts about 20 nm in width (Figs. 1.24 and 1.25).

In the successive developmental stage, the number of satellite cells increases. Only small areas of the surface of the neuronal perikaryon are not invested by satellite cells, and the cytoplasmic expansions of these cells are now more extensive, thicker, and richer in organelles than in the previous stage. Satellite cells sometimes bear solitary primary cilia lacking a central pair of microtubules. In places, adjoining satellite cells are linked by adherent junctions, gap junctions (Fig. 1.26), and short ridges (Fig. 1.26) which may represent rudiments of tight junctions (Pannese 1969; Pannese et al. 1977). Interstitial spaces (Fig. 1.26) first appear in the ganglion as intercellular clefts that are widened locally or for some length. Usually satellite cells, sometimes for a short distance undifferentiated or mitotic cells, but never nerve cells, border these primitive interstitial spaces (Fig. 1.26). These spaces usually contain patches ("interstitial bodies" of Low 1970) of a moderately dense material that at high magnification appears as a mat of thin filaments with scattered small granules. Sometimes the spaces also contain fibrils less than 10 nm thick, which may be isolated or arranged in small bundles (Fig. 1.26). A primordium of the basal lamina can be detected near the plasma

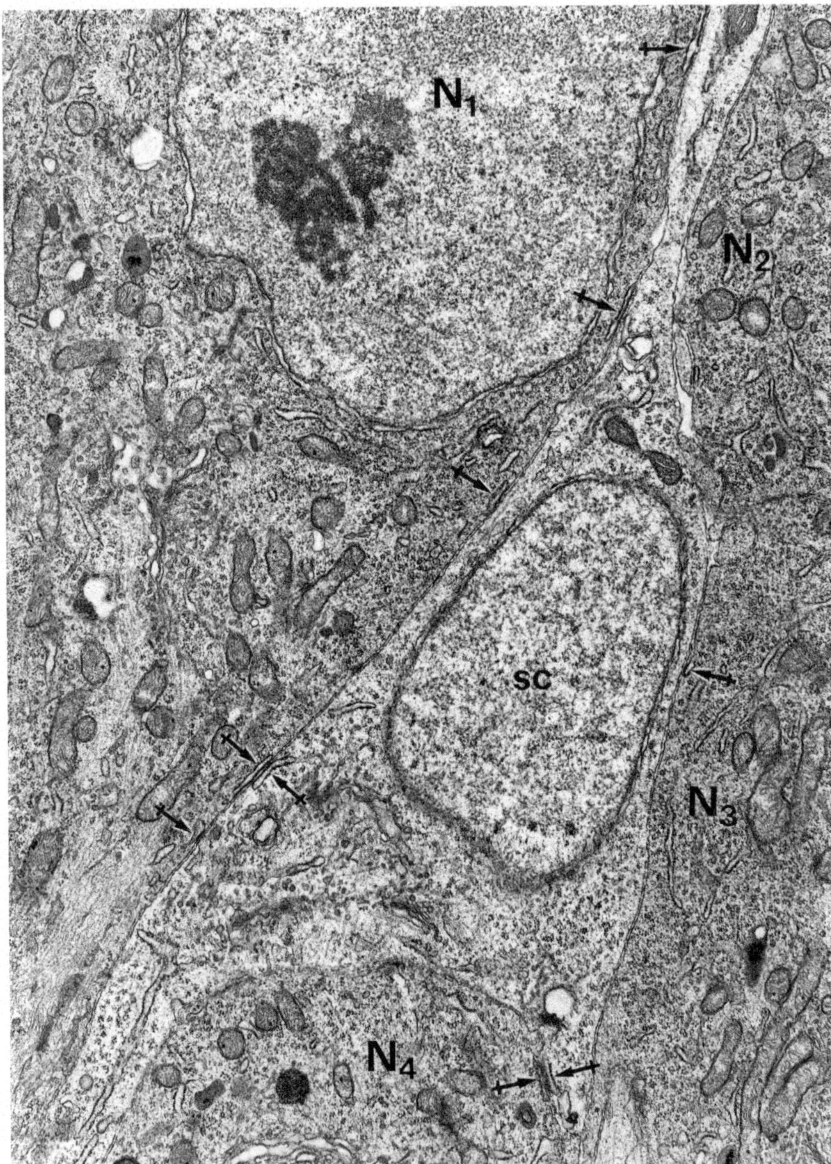

Fig. 1.25 Star shaped satellite cell (**sc**), whose body is located in the center of a group of immature neurons (**N$_1$–N$_4$**) in a chick embryo spinal ganglion. The cytoplasmic expansions of the satellite cell intervene between some of the neighboring nerve cells. Note that at this developmental stage, the neuron-satellite cell boundary is smooth. *Crossed arrows* point to subsurface cisternae; confronting subsurface cisternae can be seen between two *crossed arrows*. Transmission electron micrograph, ×15,300

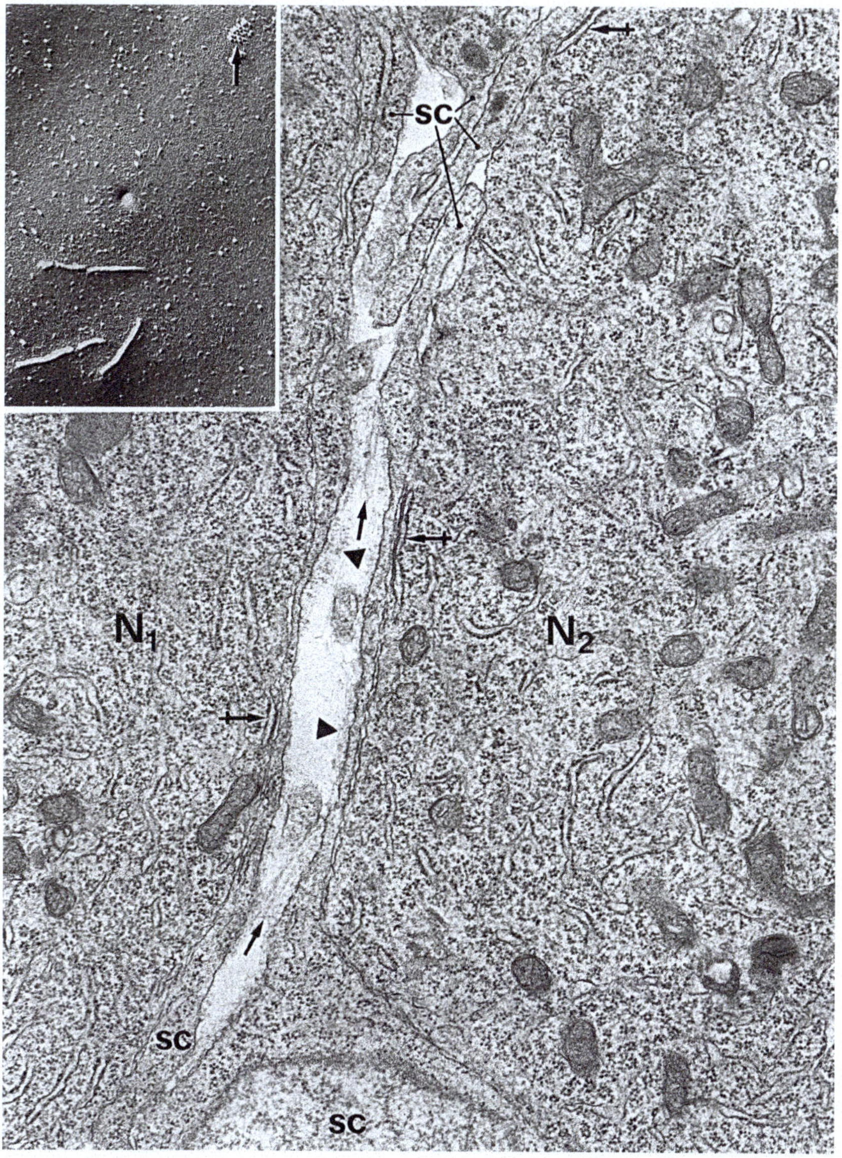

Fig. 1.26 Primitive interstitial space containing fine fibrils (*arrows*) and patches of a moderately dense material (chick embryo spinal ganglion). The interstitial space is bordered by satellite cells (**sc**) so that neurons (**N**$_1$ and **N**$_2$) do not come into direct contact with it. *Crossed arrows* point to subsurface cisternae; *black triangles* point to primordia of the basal lamina. Transmission electron micrograph, ×25,300. The P face of the split plasma membrane of a satellite cell can be seen in the inset (chick embryo spinal ganglion). A small gap junction (*arrow*) and ridges (possibly rudiments of tight junctions) are evident. Freeze-fracture replica (preparation by E. Reale, L. Luciano, and E. Pannese), ×52,400

membrane of the cells facing the primitive interstitial spaces (Fig. 1.26). At this stage, however, this lamina is not continuous, i.e., it always shows interruptions.

As development proceeds, each nerve cell comes to be completely enveloped by a satellite cell sheath. In some regions, the sheath consists of a single cell layer, while in others it consists of various overlapping cytoplasmic expansions. By this stage, the satellite cell has abandoned its earlier star shape to assume its mature flattened configuration probably by retracting some of its cytoplasmic expansions. The granular endoplasmic reticulum and the Golgi complex increase in volume. The interstitial spaces become enlarged and completely separate the satellite cell sheath related to a nerve cell from the sheaths enveloping neighboring nerve cells. As a consequence, while at the first stage each satellite cell was related to several nerve cells, at this stage each satellite cell is associated with one nerve cell only. Sometimes, more than one satellite cell appears related to a single nerve cell, so that now satellite cells outnumber nerve cells. Mitotic satellite cells are found (Pannese 1969); each belongs to a distinct perineuronal sheath, and it comes into direct contact with a single nerve cell only. The basal lamina that covers the outer contour of the satellite cell sheath still has interruptions.

At the end of body growth, satellite cells show the same general characteristics as in the preceding stage. Between these two stages, only quantitative differences can be found. In fact, in adult specimens, the satellite cell sheath is thicker and built of a greater number of cells. Different types of junction link satellite cells of each perineuronal sheath, whose outer contour is completely covered by a continuous basal lamina.

The neuron-satellite cell boundary gradually becomes more irregular and complicated as ganglion development proceeds (Pannese 1969, 1974; Yamadori 1970). When satellite cells are first detected, the outline of the nerve cells is smooth (Fig. 1.25), but this contour soon becomes irregular, on account of fingerlike projections arising from the perikaryon. These projections invaginate the apposed contour of the satellite cell or extend along the interval between the neuron and the satellite cell. Less frequently, projections from the satellite cell may also be observed. In the successive developmental stages, the surface area of the neuronal perikaryon undergoes a conspicuous increase since projections become more and more numerous and deep infoldings appear. Pinocytotic vesicles, mainly of the coated type, are frequently found in all developmental stages along both the neuronal plasma membrane and the satellite cell plasma membrane facing the neuron. Subsurface cisternae closely applied to the plasma membrane facing the satellite cells (Figs. 1.25 and 1.26) have often been observed in the nerve cells of developing sensory ganglia (Tennyson 1965, 1970; Pannese 1968a, 1974; Weis 1971; Krajčí 1973). Button-like adherent junctions may be found at the neuron-satellite cell boundary since satellite cells are detected in the ganglion (Pannese 1969; Tennyson 1970), and they persist throughout the successive developmental stages.

The boundary between the satellite cell sheath and the interstitial connective tissue, much as the neuron-satellite cell boundary, becomes gradually more irregular and complicated during ganglion development (Pannese 1969, 1974). In early development (Fig. 1.26), the plasma membrane of satellite cells is smooth or slightly wavy, so the boundary between the satellite cells and the primitive interstitial space is usually quite regular. A primordium of the basal lamina can be detected along this boundary (Fig. 1.26). Successively, projections and grooves along the outer contour of the satellite cell sheath render the boundary between the satellite cell sheath and the interstitial connective tissue increasingly irregular. Bundles of collagen fibrils may indent the surface of the satellite cell. The invaginations of this surface are sometimes deep and follow a complicated course within the satellite cell sheath.

Quantitative relationships between nerve and satellite cells change during ganglion development. Satellite cells are in fact less numerous than nerve cells in early development but come to progressively outnumber nerve cells as development proceeds (spinal ganglia: Pannese 1969; trigeminal ganglion: Lindner and Grosse 1974). The increase in number of satellite cells is due mainly to the differentiation of undifferentiated cells during early development, while later on it is chiefly due to the mitosis of fully differentiated satellite cells. The proliferative activities of both neuronal precursors and non-neuronal cells during the first half of the incubation period have been quantitatively investigated by Carr and Simpson (1978) in chick embryo spinal ganglia.

During ganglion development, the satellite cell sheath enveloping a single nerve cell body increases gradually in volume thus adjusting its total mass to the increasing neuronal size. However, due to the delay in the development of the satellite cells in respect to the neurons, the quantitative ratios between satellite and nerve cells are lower in the embryonic than in the adult ganglia.

1.12 Physiological Properties of Satellite Cells

While the basic electrophysiological properties of neurons in sensory ganglia have been carefully studied, little is known about the same properties of satellite cells. Satellite cells have often been ignored by electrophysiologists probably due to the difficulty of recording from such small cells.

The resting membrane potentials recorded from the satellite cells in the spinal ganglia of the rat (33–78 mV) and in the nodose ganglion of the vagus nerve of the cat (70–90 mV) are higher than those recorded from the corresponding neurons (35–68 and 53–54 mV, respectively) (Hösli et al. 1978; Gallego and Eyzaguirre 1978). The membrane resistance of the satellite cells of the rat spinal ganglia ranges from 4 to 42 MΩ (mean value 15.6 ± 9 MΩ), whereas that of the corresponding neurons varies from 2 to 26 MΩ (mean value 8.8 ± 6.4 MΩ) (Hösli et al. 1978). Unlike neurons, satellite cells do not give action potentials even when strong depolarizing pulses are applied (Hösli et al. 1978; Gallego and Eyzaguirre 1978).

Repetitive application of positive current pulses induces hyperpolarization of satellite cells, presumably through ejection of K^+ from the microelectrode tip, while repetitive stimulation of the vagus nerve induces slow depolarization of the same cells (nodose ganglion of the vagus nerve; Gallego and Eyzaguirre 1978). In the course of the penetration of the microelectrode in a spinal ganglion, a 10–20 mV negative potential has often been recorded immediately before the resting membrane potential of the nerve cell body has appeared (Sato and Austin 1961, in the cat). This 10–20 mV negative potential possibly corresponds to the charge of the narrow cleft between the satellite cell sheath and the nerve cell body.

A high-affinity uptake system of γ-aminobutyric acid and glutamate has been localized to satellite cells (Schon and Kelly 1974a, b; Hösli and Hösli 1978; Berger and Hediger 2000; Shoji et al. 2010). Both primary sensory neurons (de Groat 1972; de Groat et al. 1972; Feltz and Rasminsky 1974; Hösli et al. 1977) and satellite cells (Hösli et al. 1978) are depolarized by γ-aminobutyric acid. Probably this amino acid depolarizes satellite cells indirectly, inducing the release of K^+ from the neuron into the narrow cleft between neuron and satellite cells; the accumulation of K^+ induces depolarization of satellite cells (Hösli et al. 1979).

1.13 Mitotic Activity of Satellite Cells

Mitotic satellite cells (Pannese 1960, 1969; Krajčí 1975) and satellite cells labeled with tritiated thymidine (Skoglund 1967) are a frequent occurrence in sensory ganglia during body growth. Rare mitotic satellite cells (Pannese 1960; Lieberman 1976) and a small but consistent proportion of satellite cells labeled with tritiated thymidine (Friede and Johnstone 1967; Kraus-Ruppert et al. 1975 and later authors) have been observed in sensory ganglia of normal adult animals. Under normal conditions, therefore, these cells retain their ability to divide also in adult life (Fig. 1.27).

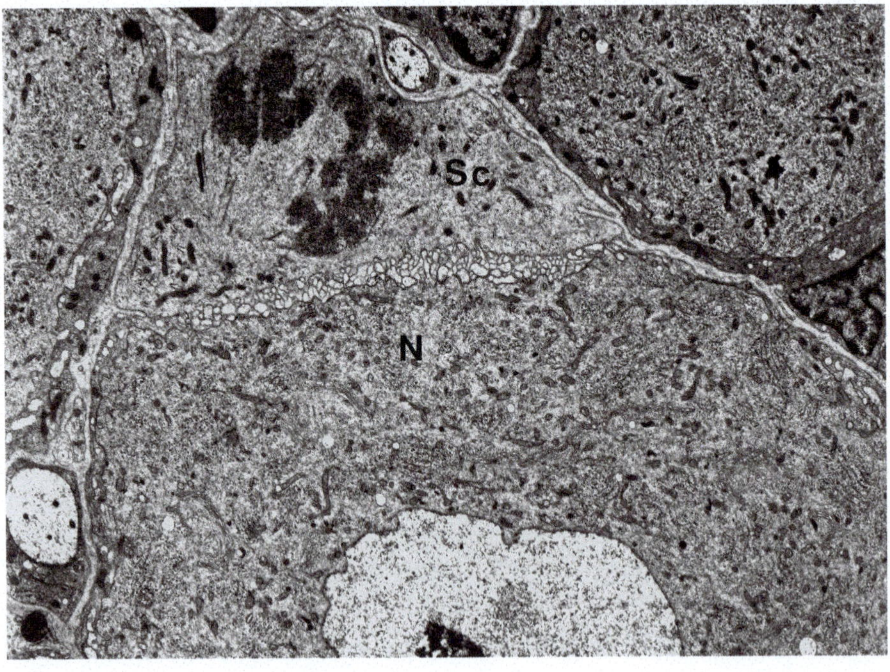

Fig. 1.27 Satellite cell (**sc**) undergoing mitosis (spinal ganglion of a normal adult rat). **N** nerve cell body. Transmission electron micrograph, ×5400

The half-life of spinal ganglion satellite cells has been estimated at about 600 days in normal adult mice (Elson et al. 2004). The question arises as to whether satellite cells arising by mitosis in the adult animal remain in situ and are added to the existing cell population, or whether they replace degenerated or otherwise lost satellite cells. Since the size of the satellite cell population in a given ganglion does not change appreciably between young adult and adult animals, it seems that new cells replace lost satellite cells in these life periods, and the satellite cell population is renewed, although at a very slow rate. For the mitotic activity of satellite cells under experimental conditions, see Sects. 1.18.1, 1.18.2, 1.18.4, and 1.18.5.

There is evidence that in animals in which growth continues throughout life, neurons of the sensory ganglia increase in number even during adult life (Devor et al. 1985; Popken and Farel 1997). New neurons could derive from the late differentiation of committed nerve cells generated prenatally (Popken and Farel 1997; Ciaroni et al. 2000). However, the possibility cannot be ruled out that new neurons derive from precursors present in the ganglion that proliferate during adult life. These precursors could be indistinguishable from perineuronal satellite cells. Therefore, some of the satellite cells that proliferate in adulthood could actually be neuronal precursors (e.g., see Kuo et al. 2005; Muratori et al. 2015).

1.14 Phagocytic Activity of Satellite Cells

Fragments of degenerated cells have been seen within perineuronal satellite cells (Fig. 1.28) both in the sensory (Pannese 1978) and autonomic ganglia (Matthews and Raisman 1972; Pilar and Landmesser 1976) of various species. These findings together with the identification of an engulfment receptor (Wu et al. 2009) on satellite cells suggest that in vivo these cells have phagocytic capacity. Satellite cells have been shown to be phagocytic when cultured in vitro (Wilkinson et al. 1999).

1.15 Age-Related Changes

As stated earlier (see Sect. 1.2), in the sensory ganglia of adult vertebrates, each nerve cell body is usually enveloped by its own satellite cell sheath. This organization does not change with age. However, both the mean volume of the satellite cell sheath and the mean value of the volume ratio between the satellite cell sheaths and the related nerve cell bodies are significantly lower in old animals than young adults (Pannese et al. 1996; Martinelli et al. 2006a). The reduction of the mean volume of the satellite cell sheaths is due in part to the significant decrease in the satellite cell number that occurs in old age (Pannese et al. 1997).

In the old age, satellite cells undergo the following structural changes. In rabbit spinal ganglia, both the total volume of the Golgi apparatus and the mean percentage

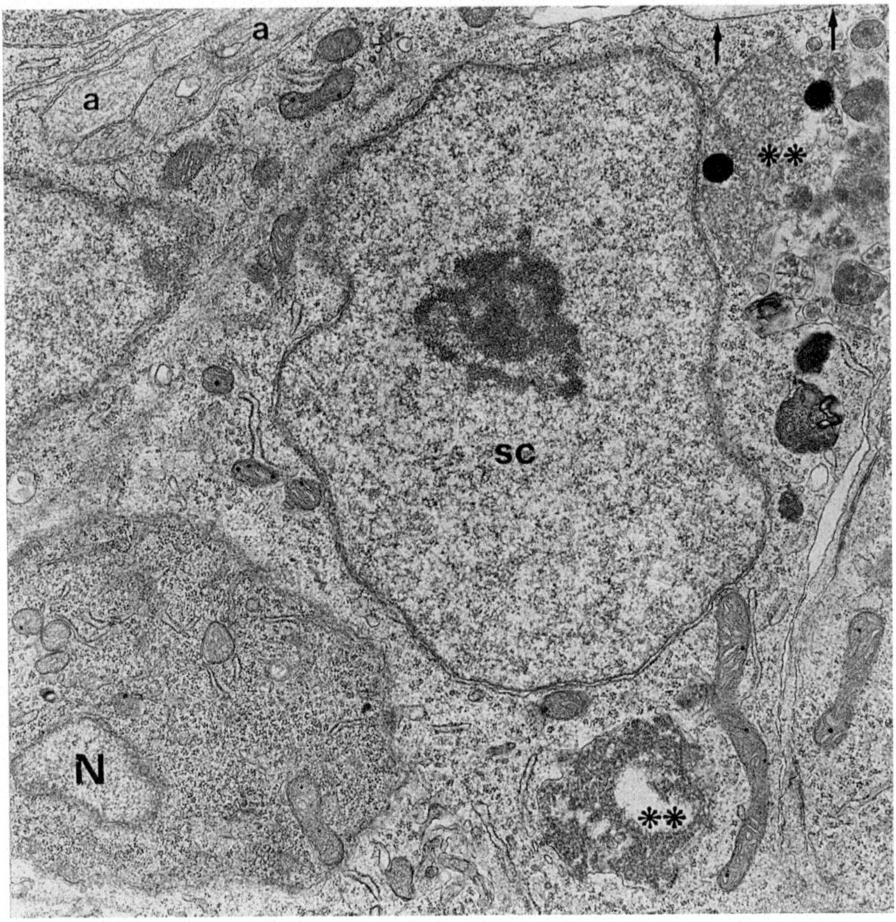

Fig. 1.28 Satellite cell (**sc**), identifiable on the basis of its relationships with a normal neuron (**N**) and a primordium of the basal lamina (*arrows*), encloses fragments (**) of degenerated nerve cells in cytoplasmic vacuoles (chick embryo spinal ganglion); **a** axon. Transmission electron micrograph, ×15,800

of cytoplasmic volume occupied by this organelle decrease significantly, while the Golgi apparatus undergoes neither structural changes nor peripheral displacement (Ledda et al. 2003). The mitochondrial structure does not change either, whereas mitochondrial size increases progressively and significantly with advancing age. Both the mitochondrial mass and the mean percentage of cytoplasmic volume occupied by mitochondria decrease progressively and significantly with age (Martinelli et al. 2007). These findings suggest that the ability of satellite cells to produce energy decreases with age and that the reduced ability of sensory neurons to respond to high energy demands in old age (e.g., see Sylvia and Rosenthal 1979) may be in part due to the diminished contribution of perineuronal satellite cells. With age, satellite cells accumulate lipofuscin (van den Bosch de Aguilar and Vanneste 1983). Lipofuscin accumulation seems to have little or no effect on either the metabolism or functional

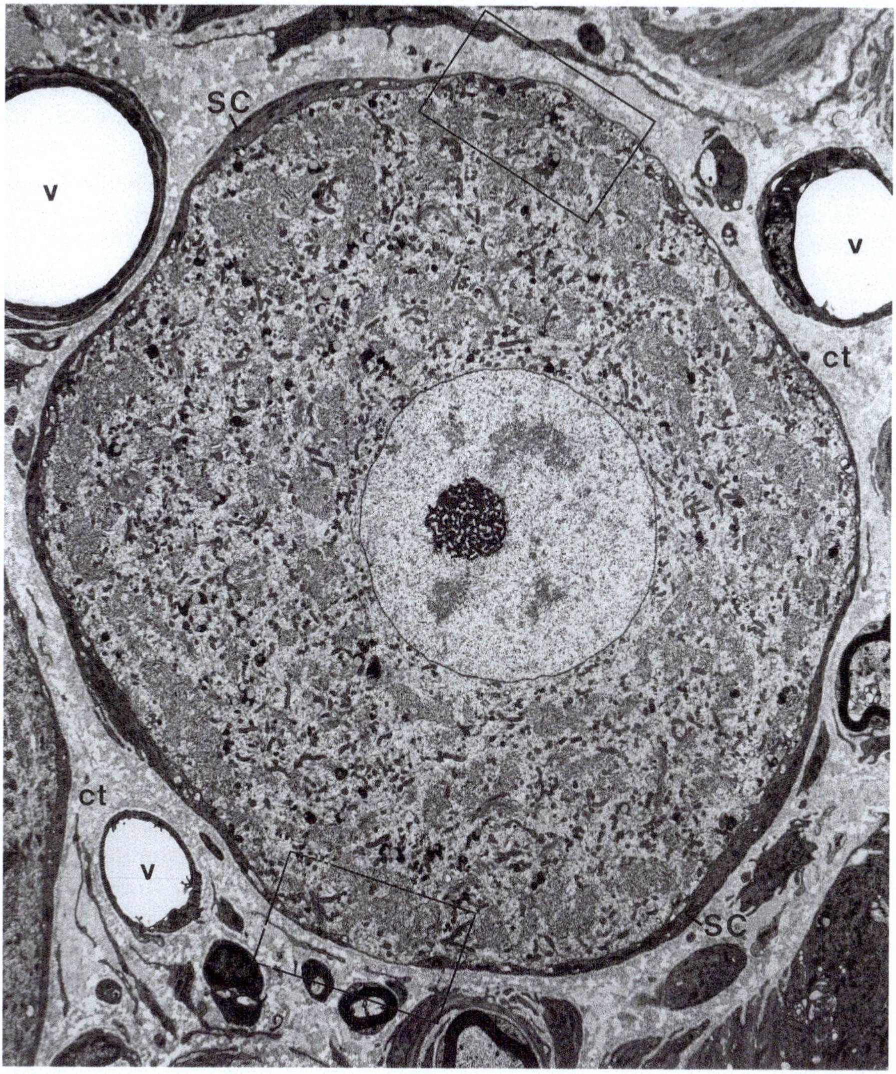

Fig. 1.29 Unit consisting of a nerve cell body and its satellite cell (**sc**) sheath (spinal ganglion of an old rabbit). Note two extensive gaps (*boxed areas*) in the satellite cell sheath; **ct** interstitial connective tissue; **v** blood vessel. Transmission electron micrograph, ×3200

activity of neurons (for a review, see Ledda et al. 1999). This is probably also the case for satellite cells.

Since satellite cells provide trophic and protective support for neurons (see Sect. 1.16), it is likely that the significant decrease in their number, the marked quantitative reduction in the volume ratio between the satellite cell sheath and the related neuron, and the decrease of the mitochondrial mass have negative consequences for neuronal activity.

Furthermore, the gaps sometimes present in satellite cell sheaths (see Sect. 1.2) are significantly more numerous and extensive in aged animals (Fig. 1.29) than young adults. For example, in rabbit spinal ganglia, for comparable neuronal perimeters, these gaps occur more than twice as often in old animals than young ones. In addition, while in young adults these gaps are no longer than 0.75 μm, in old rabbits they may reach a length of 7.7 μm (Pannese et al. 1996). Since the nerve cells in sensory ganglia lack the protection of a vascular barrier such as is present in the CNS, only the satellite cell sheath controls the traffic of substances from the blood to the ganglion neurons (see Sect. 1.16.1). Therefore, the greater number and extent of these gaps in old animals may render the neurons more susceptible to damage by harmful substances from the circulation.

The gap junctions between perineuronal satellite cells increase in number with age in rabbit spinal ganglia, whereas the mean size of individual gap junctions remains constant (Martinelli et al. 2005). Both gap junction number and dye coupling between satellite cells increase (Huang et al. 2006), whereas connexin43 expression decreases (Procacci et al. 2008) in mouse spinal ganglia in senescence. This discrepancy could depend on the presence of additional connexin types other than connexin43 in the gap junctions between satellite cells (see Sect. 1.6).

1.16 Roles of Satellite Cells Under Physiological Conditions

1.16.1 Control of Traffic to Neurons

In sensory ganglia, each neuron is usually enveloped by an individual satellite cell sheath whose outer contour faces the interstitial connective tissue containing capillaries (see Sect. 1.2). Thus, all substances from the blood must pass through the satellite cell sheath to reach the neuron. As stated above (see Sect. 1.15), neurons in sensory ganglia lack the protection of a vascular barrier such as is present in the CNS; as a consequence, it is only the satellite cell sheath that controls the traffic of substances to the ganglion neuron. Electron microscope studies using electron-opaque tracers have provided useful information on this function. The concentration of tracers within the cleft between nerve cell body and satellite cell sheath is lower than that in the connective tissue space (Rosenbluth and Wissig 1964; Arvidson 1979). Thus, the perineuronal satellite cell sheath acts as a barrier, although partially, to the movement of substances toward the neuron. Both the basal lamina and the satellite cells are involved in this function. The basal lamina acts as a filter excluding large particles and retarding the passage of other particles. After crossing the basal lamina, the tracers reach the neuronal surface by diffusing along the intercellular clefts of the sheath (Rosenbluth and Wissig 1964; Holtzman and Peterson 1969; Jacobs et al. 1976; Shinder and Devor 1994). Lanthanum penetrates the satellite cell sheath more readily at the axon hillock-initial segment than elsewhere (Shinder and Devor 1994). The chemical composition and physical state of the substance filling the intercellular clefts of the sheath, as well as fixed charges on the plasma membrane of satellite cells, may influence the diffusion of materials through the perineuronal sheath. Satellite cells might control this traffic of materials by influencing the composition and density of the substance filling the intercellular clefts and by the action of enzymes located on the plasma membranes bounding the intercellular clefts. These enzymes, whose occurrence in these sites was demonstrated by Novikoff et al. (1966), could degrade materials that might adversely affect neuronal activities. The conspicuous length and tortuosity of the pathway between the basal lamina and the neuronal surface (see Sect. 1.6) may slow down the diffusion of materials through the perineuronal sheath. The barrier function of the perineuronal satellite cell sheath may be particularly important when substances moving toward the neuron are toxic. In old age, this function may be impaired (see Sect. 1.15).

1.16.2 Homeostasis of the Perineuronal Microenvironment

Neuronal activity causes an increase in the extracellular K^+ concentration, which in turn results in an increased neuronal excitability. Satellite cells have a major role in K^+ buffering within the perineuronal microenvironment. These cells have a highly negative resting membrane potential (Hösli et al. 1978) and express inward rectifying K^+ channels (Hibino et al. 1999; Vit et al. 2006), which play a central role in the uptake of extracellular K^+. Once K^+ has entered satellite cells, it is rapidly redistributed between adjacent cells through gap junctions (Huang et al. 2005). Glutamate is the principal neurotransmitter used by sensory ganglion neurons, but high levels of glutamate are neurotoxic. If glutamate is released by sensory neurons within the perineuronal microenvironment, satellite cells, which express transporters for glutamate (Schon and Kelly 1974a, b; Hösli and Hösli 1978; Berger and Hediger 2000; Carozzi et al. 2008; Shoji et al. 2010; see Table 1.1) take up excess glutamate. In this way they maintain glutamate below neurotoxic levels in the perineuronal microenvironment. Satellite cells, that contain the enzyme glutamine synthetase (Miller et al. 2002), convert glutamate to glutamine. Glutamine is then returned to neurons and subsequently converted to glutamate by their phosphate-activated glutaminase. Neurons use glutamate for synaptic transmission or to enter their metabolic cycle. These exchanges illustrate that the metabolism of sensory ganglion neurons is highly dependent on interactions with satellite cells. γ-Aminobutyric acid can depolarize sensory ganglion neurons. Perineuronal satellite cells, which express transporters for γ-aminobutyric acid (Schon and Kelly 1974a, b; Hösli and Hösli 1978; see Table 1.1), may control the concentration of this amino acid in the perineuronal microenvironment (Young et al. 1973; Brown and Galvan 1977). Thus these cells may protect the neuronal plasma membrane from the effects of excessive concentrations of γ-aminobutyric acid—a function whose efficiency is probably improved by gap junctional coupling between satellite cells.

1.16.3 Neuroprotection

The satellite cell functions described in Sects. 1.16.1 and 1.16.2 are in themselves neuroprotective. Furthermore, in vitro studies (see Sect. 1.18.1) have shown that the presence of satellite cells in culture prolongs the survival of ganglion neurons (Shimizu 1965; Varon and Raiborn 1972; Varon et al. 1973; Mudge 1984). To explain these results, Varon (1976) suggested that satellite cells produce a trophic factor and release it to neurons. The idea that satellite cells produce neurotrophic factors and release them to the neuron with which they are associated has been invoked to explain the effects that satellite cells have on injured neurons (e.g., see Thippeswamy et al. 2005). Nitric oxide produced by injured ganglion neurons stimulates their associated satellite cells to generate cyclic guanosine 3′,5′-monophosphate. The latter seems to activate processes leading to the synthesis of neurotrophic factors in satellite cells. Neurotrophic factors released by satellite cells seem to facilitate the survival of injured neurons.

1.16.4 Influence on Neuronal Morphology

Like glial cells of the CNS (for references, see Pannese 2015), satellite cells are able to influence neuronal shape. For example, satellite cells can prevent dendrite formation on sensory neurons: when cocultured with satellite cells, neurons of the rat nodose ganglion do not grow dendrites, but in the absence of satellite cells, they do extend dendrites (De Koninck et al. 1993). Satellite cells are also able to influence the perikaryal surface of the sensory neurons with which they are associated. These cells promote, or provide a permissive environment for, the outgrowth of the perikaryal projections from neurons (Pannese et al. 1994, 1995, 1999). Furthermore, satellite cells influence neuronal maturation. Embryonic chick sensory neurons maintain a bipolar form for periods of up to 1 month if grown in vitro in the absence of non-neuronal cells; however, when glial cells of the PNS are added to the culture, the neurons change dramatically from the bipolar to the pseudounipolar form (Mudge 1984).

1.17 Communication Between Neurons and Satellite Cells

As described above, sensory ganglion neurons have numerous interactions with their associated satellite cells. However, the mechanisms of communication between these neurons and their satellite cells are poorly understood. Since gap junctions between sensory ganglion neurons and satellite cells are not present (see Sect. 1.8), the signals involved are likely to be chemical messengers. A number of studies suggest that adenosine triphosphate is the main messenger mediating neuron-to-satellite cell communication; this molecule, released from the body of sensory ganglion neurons, may activate purinergic receptors on satellite cells (Zhang et al. 2007; Gu et al. 2010).

Other molecules that seem to mediate neuron-to-satellite cell communication are nitric oxide (Morris et al. 1992), calcitonin gene-related peptide (Li et al. 2008), and endothelin-1 (Feldman-Goriachnik and Hanani 2011).

As regards the mechanisms of satellite cell-to-neuron communication, even less is known. Studies carried out on the rat nodose ganglion suggest that glutamate and γ-aminobutyric acid are involved in satellite cell-to-neuron communication (Shoji et al. 2010). Satellite cells synthesize and release prostaglandins, and it has been suggested that these molecules interact with prostaglandin receptors on adjacent neurons (Vesin et al. 1995). Finally, evidence that sensory neurons and satellite cells can communicate through bidirectional Ca^{2+} signals has been obtained in cultured mouse trigeminal ganglia (Suadicani et al. 2010). More information on satellite cell-to-neuron communication in sensory ganglia can be found in Huang et al. (2013) and Retamal et al. (2017).

1.18 Reaction of Satellite Cells to Experimental Conditions

1.18.1 In Vitro Cultures

The shape and structure of satellite cells in cultured embryonic or fetal ganglia remain similar to those in in vivo conditions even after long-term culturing (Shimizu 1965; Bunge et al. 1967; Waxman et al. 1977), whereas satellite cells undergo changes in cultures from adult animals (Belzer et al. 2010).

Neurons and satellite cells, which have been initially separated in dissociated cell cultures, show a clear tendency with time to reassociate and to reestablish their in vivo usual relationships. Waxman et al. (1977) removed spinal ganglia from chick embryos, dissociated the former into single cells, and cultivated these in vitro; they found that while in the 3-day cultures ganglion neurons usually lacked a satellite cell cover, in the 1-month cultures each neuron was enveloped by a complete satellite cell sheath. The mechanisms underlying this intercellular reaggregation may be surface recognition or attractive activities exerted by one cell type on the other. One can hypothesize that similar mechanisms are at work also during the development in vivo of sensory ganglia.

The prolonged survival in culture of sensory neurons is markedly dependent on the presence of satellite cells (Shimizu 1965; Varon et al. 1973). The presence of satellite cells in such cultures results not only in prolonged survival but also in complete maturation of ganglion neurons (Mudge 1984), in enhanced axon growth (Varon and Raiborn 1972; Varon et al. 1973), as well as in increased synthesis of neurotransmitters (Patterson and Chun 1974) and biologically active peptides (Mudge 1981). The survival in culture of the neurons from dissociated mouse spinal ganglia increases linearly with the rising of the numerical ratio between satellite cells and neurons, the highest survival of neurons being attained with a 4:1 ratio of satellite to nerve cells (Varon et al. 1974). Furthermore, in the absence of satellite cells, nerve growth factor does not improve the survival of neurons, whereas it does when at least a minimal number of satellite cells are present in the culture (Varon et al. 1974).

In cultures of sensory ganglia obtained from embryos or neonates, satellite cells proliferate (Pomerat et al. 1967; Lodin et al. 1970; Varon et al. 1973). The addition of nerve growth factor to the culture increases the mitotic activity of satellite cells (Ebendal 1975), an effect which is probably indirect, i.e., a response of satellite cells to changes in the related neurons.

Electron microscope studies have never convincingly demonstrated synaptic contacts between an axon terminal and the cell body or process of a sensory ganglion neuron in vivo. Furthermore, synaptic connections have never been found in organized cultures of sensory ganglia even after many months in culture (Bunge et al. 1967). By contrast, in dissociated cell cultures, sensory ganglion cells have been shown to develop synaptic connections (Miller et al. 1970; Lodin et al. 1973). In order to investigate influences of satellite cells on synapse formation in sensory ganglion neurons, Cooper (1984) cultivated in vitro neurons from the newborn rat nodose ganglion, with or without satellite cells; he found that synapses were numerous when neurons were cultured without satellite cells, whereas synapses were rare when neurons were cocultured with satellite cells.

The influence of satellite cells on the morphology and maturation of sensory neurons in culture is considered in Sect. 1.16.4.

1.18.2 Chemical Agents

Chemical agents have been employed in varying doses, on different species, on embryonic and adult animals, and in culture and in vivo; thus it is difficult to compare the results obtained by these various studies. Sometimes conflicting results have been obtained with the use of the same agent. For example, following administration of cadmium chloride, Tischner and Schröder (1972, in rat spinal ganglia cultured in vitro) recorded structural changes in neurons but not in satellite cells, whereas Schlaepfer (1971, in rat spinal ganglia studied in vivo) observed many tortuous and convoluted projections of satellite cells and very few alterations in nerve cells.

On the basis of the effects recorded on the nerve and satellite cells of sensory ganglia, the results can be arranged in three groups.

Some chemical agents induce similar response patterns in both neurons and satellite cells. Centrophenoxine reduces the amount of lipofuscin pigment (Meier and Glees 1971; Spoerri and Glees 1974), chlorpromazine increases the number of lysosomes (Murray and Peterson 1964; Fildes Brosnan et al. 1970), chloroquine (Tischner and Fischer 1975) and perhexiline maleate (Hauw et al. 1978) stimulate respectively the formation of lamellar bodies and of polymorphous inclusions, while some antimetabolites of thiamine (Yonezawa and Iwanami 1966) as well as compounds inducing a deficiency of pyridoxine (Yonezawa et al. 1969) produce degenerative changes. Large numbers of microtubules have been found in both spinal ganglion neurons and their associated satellite cells following exposure to taxol (Masurovsky et al. 1983).

Other chemicals produce changes in neurons without affecting satellite cells. Capsaicin induces mitochondrial swelling (Joó et al. 1969; Jancsó et al. 1977), thallium salts and sodium azide produce both mitochondrial swelling and dilatation of the cisternae of the granular endoplasmic reticulum (Hendelman 1969; Tischner and Murray 1972; Spencer et al. 1973), and phytanic acid leads to the appearance of highly osmiophilic bodies and mitochondrial alterations (Dubois-Dalcq et al. 1972). Exposure of cultured sensory ganglia to vincristine is followed by marked accumulation of neurofilaments and loss of microtubules (Journey et al. 1968), while exposure to ouabain is followed by dilatation of the Golgi complex cisternae (Whetsell and Bunge 1969). In organized cultures of mouse spinal ganglia, methadone at a low concentration determines structural changes of neurons but does not affect satellite cells, while at progressively higher concentrations, the satellite cells round up and then degenerate together with neurons (Grode and Murray 1973). This study shows that neurons are more sensitive than satellite cells to methadone and that the response of both nerve and satellite cells is related to the concentration of the tested agent. A similar dose-related response has been observed in cultures of chick spinal ganglia treated with sodium selenite (Obersteiner and Sharma 1978). On the basis of the results of these last two studies, it cannot be excluded that when administration of a chemical is followed only by changes in neurons, the absence of satellite cell response is related to the concentration of the tested agent. Possibly, changes occur also in satellite cells, but they become evident much later than in neurons and so they may escape observation. Alternatively, administration of certain chemicals might induce in satellite cells only slight metabolic changes, which do not turn into overt structural modifications.

A third group of chemical agents produce different patterns of response in neurons and in satellite cells. In cultured spinal ganglia of chick embryos, puromycin causes structural changes of the neuronal perikarya but only changes in the shape of satellite cells (Meller and Waelsch 1975), while organophosphates produce degenerative changes of nerve fibers and vacuolization of satellite cells (Watanabe and Sharma 1975). In the spinal ganglia of tricresylphosphate-poisoned slow loris, chromatolysis of the neurons and hypertrophy of the surrounding satellite cells have been observed (Ahmed 1977). The hypertrophy is characterized by cell enlargement, increase in granular endoplasmic reticulum and Golgi complex, and appearance of large numbers of ribosomes and filaments. In the sensory ganglia of rats treated with methyl mercury compounds, the following changes have been observed: neuronal chromatolysis and increase in the number and size of the satellite cells surrounding the affected neurons (methyl mercury hydroxide; Herman et al. 1973); displacement of membrane-associated ribosomes in the neuronal perikaryon, structural changes of the neuronal nucleus, axon degeneration, and a proliferative response of satellite cells (methyl mercury chloride; Jacobs et al. 1975); shrinkage of the nerve cell body, vacuolization of the neuronal mitochondria, and proliferation of the satellite cell microvilli (methyl mercury chloride and mercuric bichloride; Chang and Hartmann 1972). Treatment with cisplatin also induces differing changes in neurons and in satellite cells of rat sensory ganglia (Cece et al. 1995; Corsetti et al. 2000; Sugimoto et al. 2001). In spinal ganglia of the rabbit poisoned with carbon tetrachloride, neurons undergo degenerative changes, while satellite cells increase in number (Lahl 1975); the neuronal degeneration is followed by formation of Nageotte nodules (see Sect. 1.19.3). Marked numbers of Nageotte nodules have also been found in the spinal ganglia of rats treated with paclitaxel (Peters et al. 2007). In the spinal ganglia of rats treated with lead acetate, some neurons display an increase in neurofilaments and a relative paucity of endoplasmic reticulum, while satellite cells proliferate and numerous dense bodies appear in their cytoplasm (Schlaepfer 1969). In the spinal ganglia of rats treated with isoniazid, neurons develop structural changes resembling those occurring following axon injury, whereas the satellite cells enveloping the damaged neurons undergo mitotic division (Schröder 1970). The changes undergone by satellite cells in many of the above experiments seem to reflect increased cellular activity. These changes do not seem to be due to a direct effect of the chemical substances on satellite cells but rather the result of a response of these cells to the changes occurring in the related neurons.

Although it is difficult to draw definite conclusions, the available data suggest that neurons of sensory ganglia are more sensitive than satellite cells to many chemical agents. This provisional conclusion is supported by the following observations: some chemicals produce changes in neurons without visibly affecting satellite cells; other chemicals induce more severe changes in neurons than in satellite cells; changes in neurons are caused by lower doses of certain chemical agents than those inducing changes in satellite cells; finally, no chemical agent so far tested produces changes in satellite cells leaving neurons unaffected.

1.18.3 Radiation

Data on this topic are scarce. In organized cultures of rat spinal ganglia, shortly after X-irradiation many satellite cells undergo acute degeneration, leaving only the basal lamina covering the neurons (Masurovsky et al. 1967). At first, neurons display limited changes, but later they develop a number of prominent alterations including chromatolysis, cytoplasmic vacuolization, formation of autophagic vacuoles, and mitochondrial changes. With time, some neurons undergo terminal degeneration. These results indicate that satellite cells are highly sensitive to X-irradiation. By contrast, satellite cells are relatively resistant to irradiation with protons. After proton irradiation most of the satellite cells of rat spinal ganglia retain their normal structure, some display lipid droplets and lipofuscine granules, while only a few undergo degeneration (Andres 1963; Andres et al. 1963). The satellite cells of autonomic ganglia also undergo structural changes following X-irradiation (Forssmann et al. 1966; Chang et al. 1973).

1.18.4 Axon Injury

Following axon injury of the neuron with which they are associated, satellite cells increase in number and undergo a series of modifications. The increase in number of satellite cells occurs by mitotic division of the existing satellite cells

(Pannese 1964; Friede and Johnstone 1967; Leech 1967; Humbertson et al. 1969; Lu and Richardson 1991; Shinder et al. 1999; Fenzi et al. 2001; Zhang et al. 2007; Donegan et al. 2013; Christie et al. 2015; Muratori et al. 2015). The increase in surface area of the nerve cell body (Zimmerman et al. 1971) or a mitogenic protein released by injured neurons (Wen et al. 1994) could be signals that trigger satellite cell proliferation. Once increased in number, satellite cells might more adequately meet the metabolic requirements of the neuron, which increase during axon regeneration (see Lieberman 1971 for references). The modifications of satellite cells include structural changes involving nucleolus, Golgi apparatus, endoplasmic reticulum, mitochondria, and dense bodies (Cervós-Navarro 1962; Pannese 1964; Nathaniel and Nathaniel 1973), changes in the activity of some enzymes (Schwandt 1976; Harvarik 1977; Glover 1982; Magnusson et al. 2000), and the increased expression of several molecules, e.g., glial fibrillary acidic protein (Elfvin et al. 1987; Woodham et al. 1989; Stephenson and Byers 1995; Chudler et al. 1997; Vit et al. 2006; Takeda et al. 2007), S100 proteins (Sandelin et al. 2004; Levy et al. 2007), transforming growth factors (Xian and Zhou 1999), fibroblast growth factors (Grothe et al. 1997), receptors to neurotrophins (Zhou et al. 1996), tumor necrosis factor-α (Ohtori et al. 2004), cell adhesion molecules (Zhang et al. 2000), extracellular signal-regulated protein kinase (Rashid et al. 2004), interleukin 1β (Takeda et al. 2007), and major histocompatibility complex class II antigens (Gehrmann et al. 1991).

Further changes occurring following axon injury include the formation of bridges connecting previously separate perineuronal sheaths (Fig. 1.30) and a marked increase in number of the gap junctions between satellite cells (Pannese et al. 2003). While in normal conditions satellite cells are only coupled to other satellite cells within the same perineuronal sheath (see Sect. 1.6), following axon injury satellite cells become extensively coupled to satellite cells belonging to other perineuronal sheaths. The marked increase in number of gap junctions correlates with the increase in the incidence of dye coupling among satellite cells (Hanani et al. 2002). These changes could provide the basis for medium- and long-distance communication between satellite cells. Changes occurring in satellite cells following axon injury to the associated ganglion neuron show that these cells are endowed with marked plasticity. Both glial fibrillary acidic protein within satellite cells and coupling through gap junctions between these cells also increase markedly following compression injury of the ganglion (Zhang et al. 2009).

The number of sympathetic nerve fibers surrounding nerve cell bodies in spinal ganglia (see Sect. 1.8) increases notably after axon injury (McLachlan et al. 1993; Shinder et al. 1999). Some of these fibers are closely associated with satellite cells, that contain nerve growth factor and neurotrophin 3 (Zhou et al. 1999). Blocking the binding of these growth factors by specific antibodies reduces the sympathetic sprouting, indicating that growth factors released by satellite cells are major contributors to sympathetic sprouting (Zhou et al. 1999). It has been suggested that this sprouting contributes to the generation and maintenance of some forms of neuropathic pain; however, convincing data are lacking. Sciatic nerve crush also causes upregulation of cell adhesion molecules by some sensory neurons and satellite cells (Zhang et al. 2000), giving rise to the suggestion that these molecules contribute to sympathetic nerve sprouting.

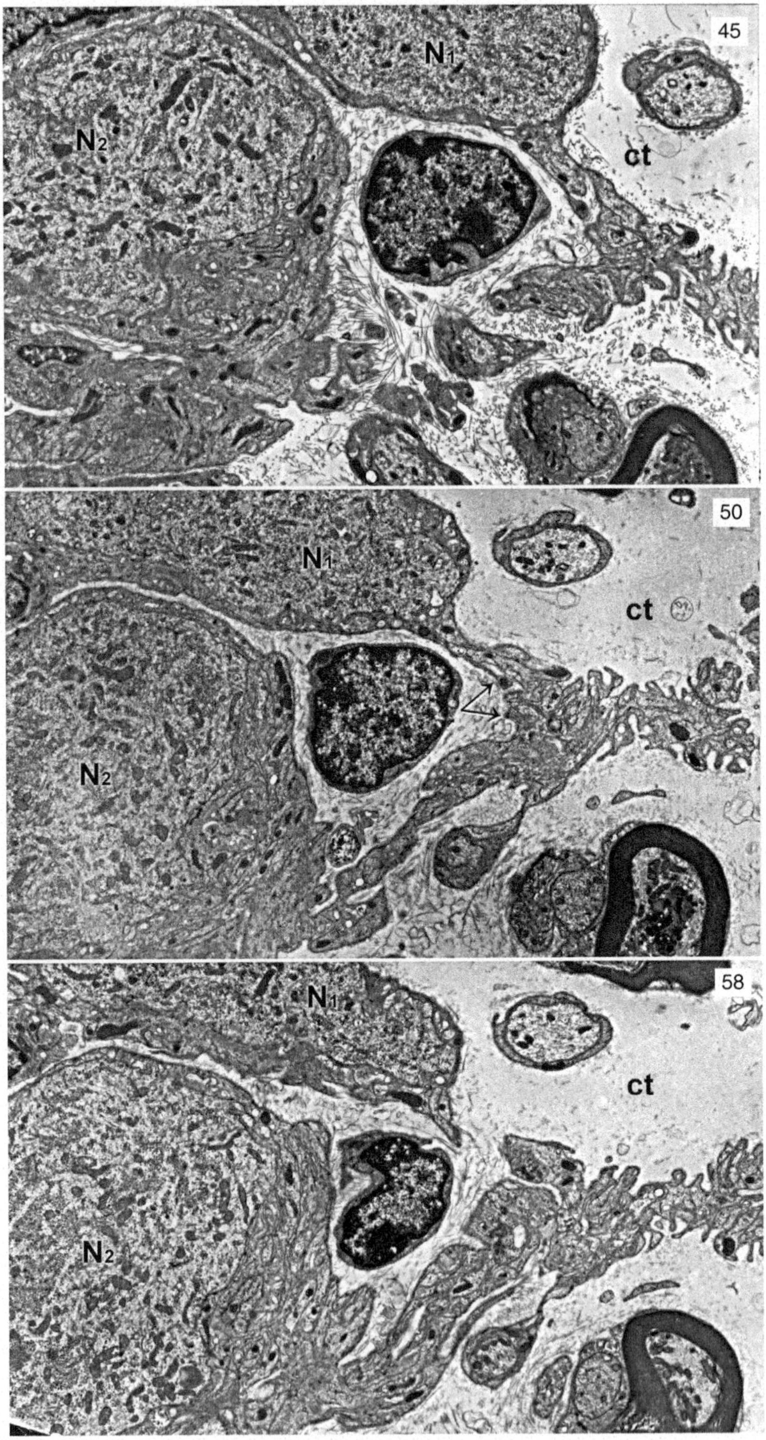
45
N1
N2
ct
50
N1
ct
N2
58
N1
ct
N2

1.18.5 Other Experimental Conditions

When the nerve cell bodies in sensory ganglia hypertrophy as result of the enlargement of their peripheral field of innervation, the satellite cells increase in number due to increased proliferation and undergo structural changes involving principally the nucleolus and Golgi apparatus (Pannese 1964). Similar findings have been recorded in autonomic ganglia (Yarygin et al. 1976). The rise in number of satellite cells enveloping a hypertrophied nerve cell body again emphasizes the existence in sensory ganglia of a quantitative balance between the volume of the nerve cell body and that of its associated satellite cell sheath (see Sect. 1.10).

In sensory ganglia of ascorbic acid-deficient guinea pigs, breakdown of Nissl substance, loss of neurofilaments and microtubules, and swelling of mitochondria have been observed in neurons, while dilatation of granular endoplasmic reticulum cisternae has been observed in satellite cells. Furthermore, some neurons have been found deprived of the satellite cell sheath, their surface lying directly under the basal lamina (Sulkin et al. 1973). Satellite cell proliferation has been observed in spinal ganglia of vitamin E-deficient rats (Cecchini et al. 1999).

The findings summarized above show that neurons and satellite cells differ in their individual sensitivity to many experimental conditions and that damage to one of these two cell types influences the behavior of the other, so stressing a close interaction between the two.

1.19 Reaction of Satellite Cells to Pathological Conditions

1.19.1 Viral Infections

Herpes simplex infection is the most studied of the viral infections affecting sensory ganglia. During initial infection, virus travels by retrograde axonal transport from the skin to the bodies of sensory ganglion neurons in which it establishes a latent infection. Sensory ganglion neurons are capable of supporting the complete replication of the virus with the production of mature enveloped particles; by contrast, herpes simplex infection is rarely detected in satellite cells in vivo, and these cells contain only unenveloped virus particles (Dillard et al. 1972; Cook and Stevens 1973; Hill and Field 1973). In vivo, sensory ganglion neurons usually show the following structural changes associated with herpes simplex infection: margination of chromatin, disruption of the nucleolus, and reduplication of the nuclear membrane

Fig. 1.30 Three sections from a series showing the relationships between the satellite cell sheaths enveloping two distinct neurons (**N**$_1$ and **N**$_2$) after axon injury (mouse spinal ganglion). Note that the two perineuronal sheaths are connected by a bridge (*arrowed* in section 50). The numbering of each micrograph indicates the corresponding section in the series; **ct** interstitial connective tissue. Transmission electron micrographs, ×4300

(Dillard et al. 1972; Cook and Stevens 1973; Hill and Field 1973; Whetsell et al. 1977; Ziegler and Pozos 1977). Structural changes to perikaryal organelles are observed less frequently (Ziegler and Pozos 1977). Satellite cells infected with herpes simplex virus usually show margination of the chromatin and reduplication of the nuclear membrane (Dillard et al. 1972; Cook and Stevens 1973; Hill and Field 1973; Ziegler and Pozos 1977); less frequently they show dilatation of the endoplasmic reticulum cisternae and mitochondrial changes (Ziegler and Pozos 1977). In vivo neurons infected with herpes simplex survive, while in culture they are destroyed by the virus (Ziegler and Herman 1980). Infected neurons are usually found dispersed within a ganglion. The virus does not spread from infected neurons to adjacent neurons probably because the satellite cells form a barrier to the interneuronal virus spread within the ganglion (LaVail et al. 1997; Wilkinson et al. 1999). In the vestibular ganglion, virus particles cannot spread from neurons to satellite cells because they cannot pass through the loose myelin sheath (Wakisaka et al. 2001). In sensory ganglia of experimentally infected mice, satellite cells proliferate and the substantial increase in number of these cells probably facilitates neuronal survival (Elson et al. 2003).

In acute varicella-zoster infection of sensory ganglia, both neurons and satellite cells contain viral particles (Reichelt et al. 2008; Zerboni and Arvin 2015; Fig. 1.31). In the latent phase, virus particles are localized in the neurons of sensory ganglia but rarely in satellite cells (Levin et al. 2003). Varicella-zoster virus can induce cell-cell fusion and polykaryon formation between a neuron and its enveloping satellite cells (Reichelt et al. 2008).

Complete replications of the pseudorabies virus have been observed with the electron microscope within neurons, while defective replications have been observed in supporting cells of sensory ganglia (Field and Hill 1974). Large numbers of viral particles have been found in both neurons and satellite cells of the sensory ganglia of rabbits infected with Aujeszky's disease virus; however, whereas evidence of complete viral replication within the neurons has been obtained, only defective replication seems to occur in satellite cells (Becker 1968). By contrast, in the sensory ganglia of calves infected with the same virus, the neurons contain large numbers of viral particles, but infection does not spread to satellite cells, not even those which ensheath an infected neuron (McCracken and Dow 1973b). In pigs infected with swine vesicular disease virus, satellite cells of sensory and autonomic ganglia show intranuclear inclusion bodies, but no viral particles have been found within them (Lenghaus et al. 1976).

1.19.2 Hereditary Neuropathies

In the sensory ganglia of mice affected by dystonia musculorum, some neurons contain large bundles of neurofilaments, whereas only minor structural changes have been occasionally observed in satellite cells (Janota 1972). In the sensory ganglia of the Sprawling mouse and of the mutilated foot rat, many nerve cells show structural changes similar to those seen in chromatolytic neurons (Fig. 1.32), whereas satellite cells apparently remain unmodified (Duchen and Scaravilli 1977).

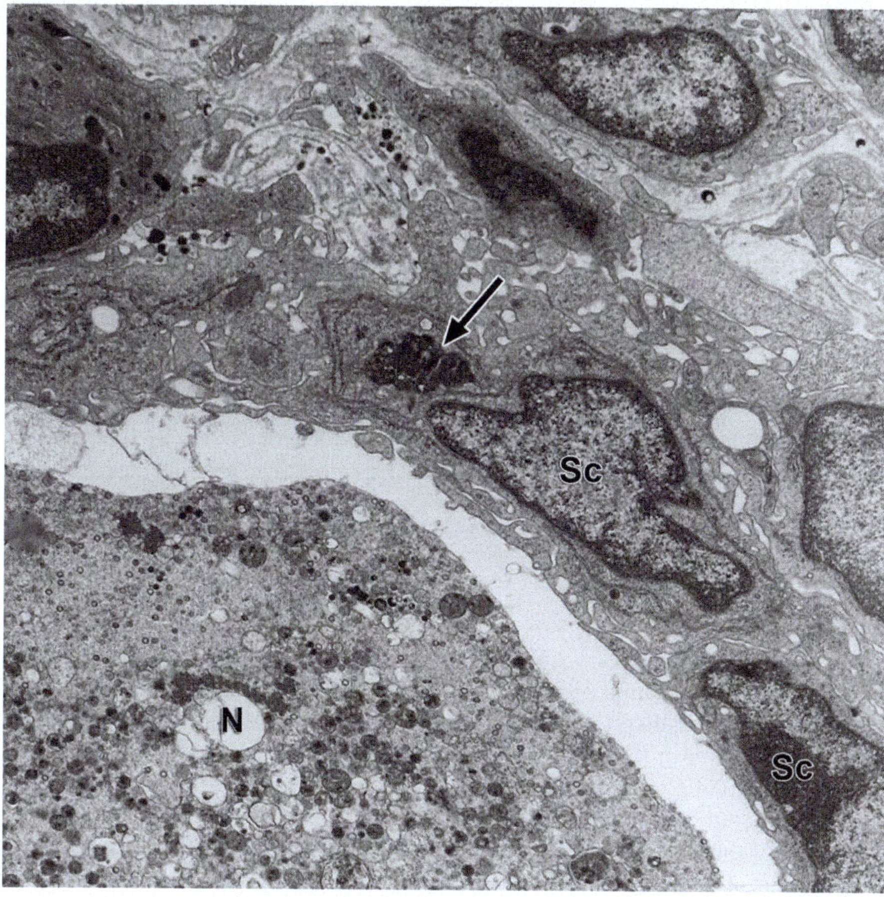

Fig. 1.31 Nerve cell body (**N**) and its satellite cells (**Sc**) containing virions with characteristics of varicella-zoster virus particles. Most varicella-zoster virions within neuron and satellite cells are unenveloped and lack an electron-dense core. A satellite cell at a very late stage of infection with a large viral replication center is arrowed (varicella-zoster-virus-infected human spinal ganglion). Transmission electron micrograph ×5800 (courtesy of L. Zerboni)

In Friedreich's ataxia, the satellite cells of sensory ganglia undergo proliferation and are affected by iron dysmetabolism (Koeppen et al. 2009). In familial amyloidotic polyneuropathy, in which mutated transthyretin accumulates in the extracellular space of spinal ganglia, satellite cells are able to uptake and degrade this protein, thus contributing to its clearance (Gonçalves et al. 2014).

1.19.3 Degenerative Disorders

Various chronic degenerative disorders result in loss of ganglionic neurons (see reviews in De Castro 1932; Prineas and Spencer 1975). Satellite cell proliferation appears to be consistently associated with degeneration of ganglion neurons.

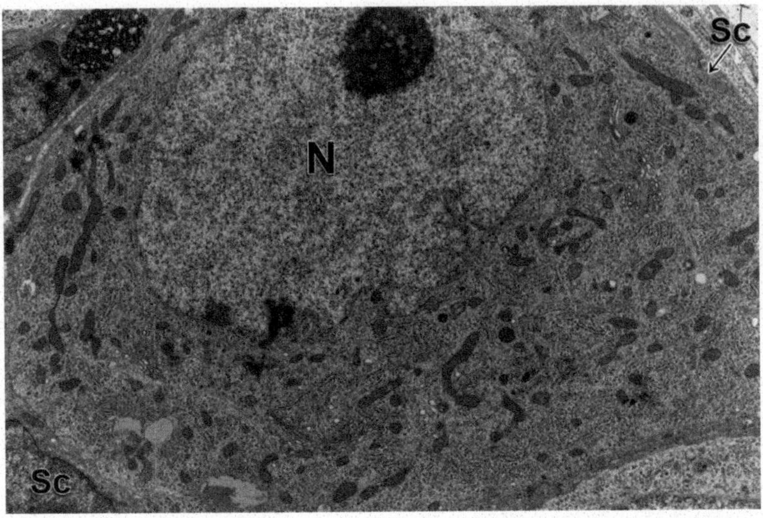

Fig. 1.32 Nerve cell body (**N**) with eccentric nucleus and wrinkled nuclear surface (spinal ganglion of a mutilated foot rat). Unlike the nerve cell body, satellite cells (**Sc**) appear unmodified. Transmission electron micrograph ×4500 (Courtesy of F. Scaravilli)

This proliferation does not seem to be a direct effect of the pathological injury on satellite cells, but rather a secondary response of these cells to the degenerative changes occurring in the neurons to which they are associated. In these conditions, satellite cells probably participate in the elimination of degenerated neurons. A Nageotte nodule (Nageotte 1907a, b) may remain in the area formerly occupied by the degenerated nerve cell body. The nodule, also called residual nodule (Figs. 1.33 and 1.34), is a cluster of satellite cells encircled by a connective tissue envelope.

1.20 Role of Satellite Cells in Neuropathic Pain

Peripheral nerve injury and inflammation of peripheral tissues increase the excitability of sensory ganglion neurons (Wall and Devor 1983; Zhang et al. 1997; Cherkas et al. 2004), causing them to become spontaneously active or discharge at a much lower threshold than normal. This is a major factor in the generation and maintenance of neuropathic pain (Devor and Seltzer 1999). As described earlier (see Sect. 1.2), in the sensory ganglia of adult vertebrates, each nerve cell body is tightly enveloped by an individual satellite cell sheath, which is in turn completely surrounded by connective tissue to form a discrete unit (Pannese 1981). Therefore, satellite cells are the only cells to be in direct contact with the body of the sensory neuron. Since satellite cells are highly sensitive to changes in neuronal state and are

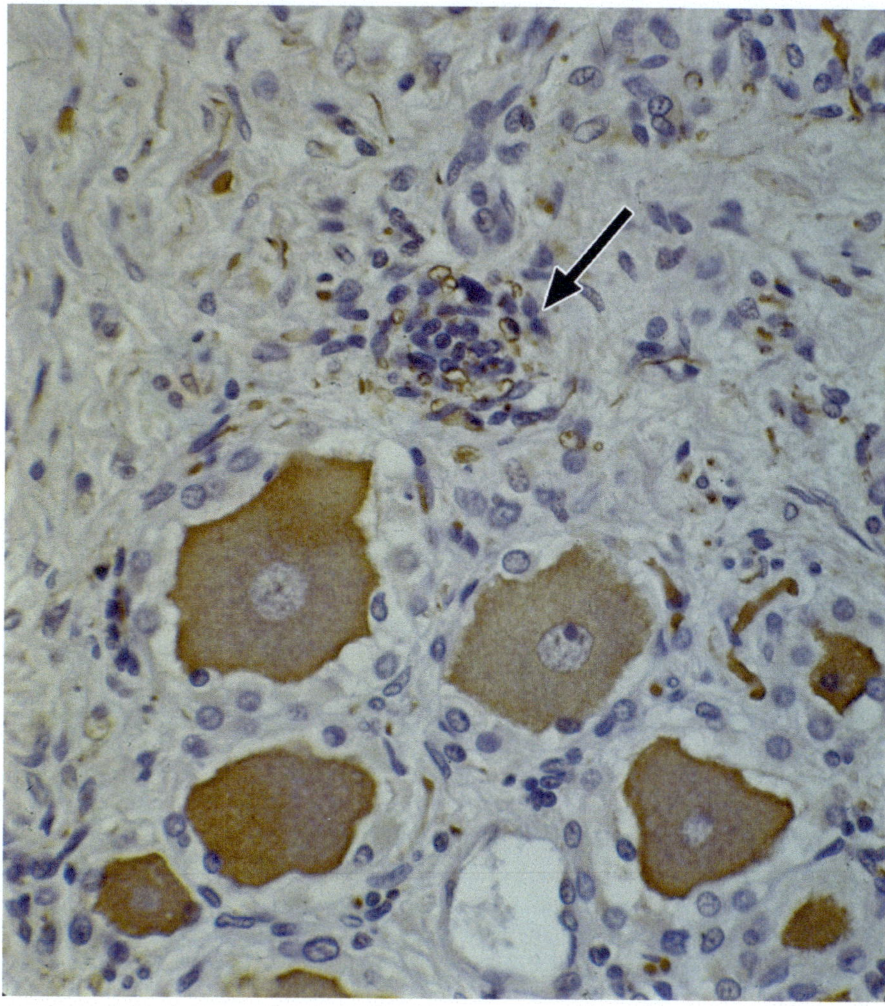

Fig. 1.33 Nerve cell bodies with their satellite cells (human spinal ganglion). *Arrow* points to a Nageotte nodule. Light micrograph of a section immunostained for neurofilament proteins; the nuclei are stained with hematoxylin ×460 (Courtesy of F. Scaravilli)

able to modulate neuronal excitability (see Sect. 1.16.2), they are thought to play an important role in the development and maintenance of neuropathic pain. Several mechanisms underlying the contribution of satellite cells to the hyperexcitability of sensory ganglion neurons, and hence the development and maintenance of neuropathic pain, have been proposed. Some of these proposed mechanisms are here summarized. Since neurons are the lesion target, satellite cell changes occur secondary to the neuronal changes, after the activation of signaling mechanisms between injured neurons and associated satellite cells. Adenosine triphosphate

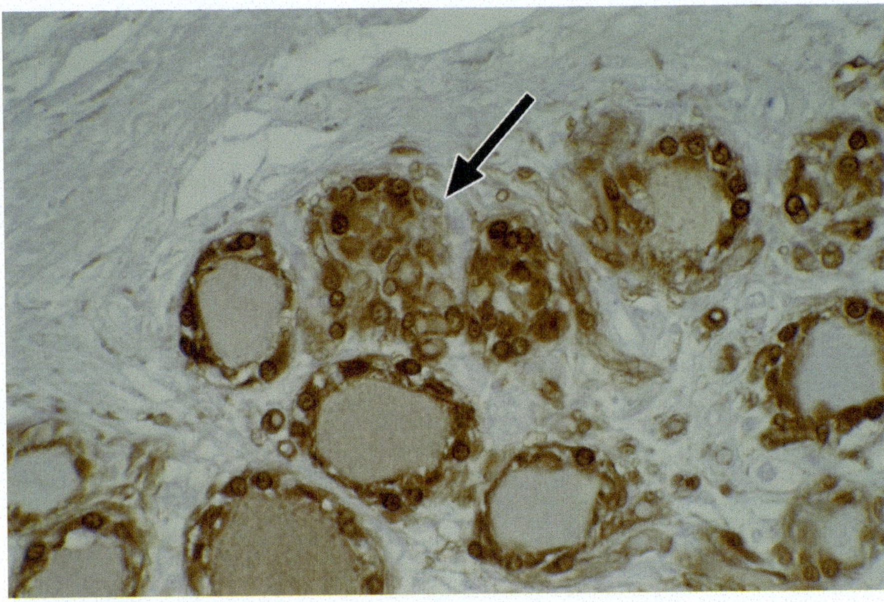

Fig. 1.34 Nerve cell bodies surrounded by their satellite cells expressing S100 proteins (human spinal ganglion). *Arrow* points to a Nageotte nodule. Light micrograph ×340 (Courtesy of F. Scaravilli)

released by neurons may activate purinergic receptors on satellite cells, leading to the synthesis of cytokines (e.g., interleukin 1β and tumor necrosis factor α) by satellite cells (Ohtori et al. 2004; Li et al. 2005; Dubový et al. 2010). The cytokines thus released may in turn enhance neuronal excitability (Takeda et al. 2007). Evidence in support of this hypothesis was obtained by Takeda et al. (2008): the local application of an antagonist of the interleukin receptor 1 restores the physiological threshold of sensory ganglion neurons and reduces the pain associated with peripheral inflammation.

Peripheral nerve injury and facial inflammation increase the excitability of trigeminal ganglion neurons. In such conditions, these neurons release K^+ to the perineuronal microenvironment. The concentration of K^+ within this space is primarily regulated by the K^+ channel Kir4.1 expressed by satellite cells. However, peripheral nerve injury and facial inflammation downregulate Kir4.1 channel in satellite cells (Vit et al. 2008; Takeda et al. 2011) with consequent impairment of K^+ homeostasis within the perineuronal microenvironment. The increase in extracellular K^+ leads to a further increase in neuronal excitability, which may further worsen neuropathic pain. Furthermore, Kir4.1 silencing in satellite cells using RNA interference results in pain states in the absence of nerve injury (Vit et al. 2008). Taken together, these findings suggest that Kir4.1 channel expressed by satellite cells is involved in the development and maintenance of neuropathic pain. Since Kir4.1 is also involved in glutamate homeostasis (Djukic et al. 2007), the satellite cell Kir4.1 channel could contribute to neuropathic pain by an additional mechanism. If glutamate—the principal neurotransmitter of sensory ganglion neurons—is released into the perineuronal microenvironment, the impaired glutamate homeostasis caused by reduction in satellite cell Kir4.1 could lead to an increase in extracellular glutamate and as a consequence an increase of neuronal excitability.

Gap junctions between satellite cells increase in number following nerve injury (Hanani et al. 2002; Pannese et al. 2003). The increased coupling between satellite cells could cause alterations in the perineuronal microenvironment (Ohara et al. 2009) so contributing to generating or maintaining neuropathic pain (e.g., see Hanani et al. 2002; Cherkas et al. 2004). Evidence in support of this hypothesis is as follows: gap junction blockers restore neuronal excitability to control levels and reduce pain in various pain models (e.g., see Spataro et al. 2004; Dublin and Hanani 2007; Huang et al. 2010; Jasmin et al. 2010). Furthermore, by using RNA interference to neutralize connexin43—the major structural component of gap junctions expressed by satellite cells (Procacci et al. 2008)—in a model of facial neuropathic pain, Ohara et al. (2008) reduced the number of gap junctions in the trigeminal ganglion and produced analgesia.

After nerve injury or inflammation, neurotrophic factors increase the excitability of sensory ganglion neurons and thus can contribute to neuropathic pain (for more details, see Julius and Basbaum 2001). Satellite cells express both neurotrophic factors (Wetmore and Olson 1995; Hammarberg et al. 1996; Gill and Windebank 1998; Zhou et al. 1999) and their receptors (Zimmermann and Sutter 1983; Wetmore and Olson 1995; Zhou et al. 1996; Pannese and Procacci 2002) and upregulate both following axon injury (Vaegter 2014). Thus, we have an additional mechanism by which satellite cells may be involved in the generation and maintenance of neuropathic pain. This suggestion is supported by the finding that treatment of animals with antibodies against receptors to neurotrophic factors reduces pain (Kobayashi et al. 2015). The association between neuropathic pain and sympathetic fibers surrounding the nerve cell bodies of spinal ganglia is discussed in Sect. 1.18.4. Hypotheses based on the existence of gap junctions between satellite cells and sensory ganglion neurons are not discussed since, as far as the author is aware, gap junctions have never been described at that site (see Sect. 1.8).

The effects of peripheral nerve injury may spread from directly affected neurons to unaffected ones (Stephenson and Byers 1995). As described in Sect. 1.18.4 and shown in Fig. 1.30, following injury to the axon of the neuron with which they are associated, satellite cells form bridges to adjacent and previously separate perineuronal sheaths (Hanani et al. 2002; Pannese et al. 2003). These bridges may provide the basis for the propagation of signals from morpho-functional units affected by peripheral injury to unaffected units.

Numerous studies on the relationships between glia and pain have been carried out over the past decade or so. They have established that perineuronal satellite cells undergo various changes when neuropathic pain occurs. Each of these changes has inspired a particular hypothesis as to the mechanisms by which satellite cells contribute to the development and maintenance of neuropathic pain. However, decisive experiments to confirm or refute these hypotheses are lacking.

References

Adamo NJ, Daigneault EA (1972) Desmosome-like junctions in the spiral ganglia of cats. Am J Anat 135:141–146

Adamo NJ, Daigneault EA (1973) Ultrastructural features of neurons and nerve fibres in the spiral ganglia of cats. J Neurocytol 2:91–103

Adoutte A, Balmefrézol M, Beisson J, André J (1972) The effects of erythromycin and chloramphenicol on the ultrastructure of mitochondria in sensitive and resistant strains of Paramecium. J Cell Biol 54:8–19

Ahmed MM (1977) Changes in the ultrastructure of satellite cells of slow loris in tricresylphosphate poisoning. Acta Neuropathol 37:173–175

Ajima H, Kawano Y, Takagi R, Aita M, Gomi H, Byers MR, Maeda T (2001) The exact expression of glial fibrillary acidic protein (GFAP) in trigeminal ganglion and dental pulp. Arch Histol Cytol 64:503–511

Alvarez MP, Solas MT, Suarez I, Fernandez B (1989) Glial fibrillary acidic protein-like immunoreactivity in cat satellite cells of sympathetic ganglia. Acta Anat 136:9–11

Andres KH (1961) Untersuchungen über den Feinbau von Spinalganglien. Z Zellforsch 55:1–48

Andres KH (1963) Elektronenmikroskopische Untersuchungen über Strukturveränderungen im Zytoplasma von Spinalganglienzellen der Ratte nach Bestrahlung mit 185 MeV-Protonen. Z Zellforsch 60:633–658

Andres KH, Larsson B, Rexed B (1963) Zur Morphogenese der akuten Strahlenschädigung in Rattenspinalganglien nach Bestrahlung mit 185 MeV-Protonen. Z Zellforsch 60:532–559

Anzil AP, Blinzinger K, Herrlinger H (1976) Fenestrated blood capillaries in rat cranio-spinal sensory ganglia. Cell Tissue Res 167:563–567

Aoki E, Semba R, Kashiwamata S (1991) Evidence for the presence of L-arginine in the glial components of the peripheral nervous system. Brain Res 559:159–162

Aoki E, Takeuchi IK, Shoji R, Semba R (1993) Localization of nitric oxide-related substances in the peripheral nervous tissues. Brain Res 620:142–145

Arnold W (1970) Ungewöhnlich grosse sphärische Lipidkörper im Ependym und Subependym des Feuersalamanders. Z Zellforsch 106:523–538

Arvidson B (1979) Distribution of intravenously injected protein tracers in peripheral ganglia of adult mice. Exp Neurol 63:388–410

Averill S, Delcroix J-D, Michael GJ, Tomlinson DR, Fernyhough P, Priestley JV (2001) Nerve growth factor modulates the activation status and fast axonal transport of ERK ½ in adult nociceptive neurones. Mol Cell Neurosci 18:183–196

E. Pannese, *Biology and Pathology of Perineuronal Satellite Cells in Sensory Ganglia*, Advances in Anatomy, Embryology and Cell Biology 226,
https://doi.org/10.1007/978-3-319-60140-3

Bär K-J, Schurigt U, Scholze A, Segond von Banchet G, Stopfel N, Bräuer R, Halbhuber K-J, Schaible H-G (2004) The expression and localization of somatostatin receptors in dorsal root ganglion neurons of normal and monoarthritic rats. Neuroscience 127:197–206

Becker C-H (1968) Die Multiplikation des Aujeszkyschen Virus in den Spinalganglien des Kaninchens. Arch Exp Veterinaermed 22:363–381

Belzer V, Shraer N, Hanani M (2010) Phenotypic changes in satellite glial cells in cultured trigeminal ganglia. Neuron Glia Biol 6:237–243

Bennett G, Hemming R (1989) Ultrastructural localization of CMPase, TPPase, and NADPase activity in neurons, satellite cells, and Schwann cells in frog dorsal root ganglia. J Histochem Cytochem 37:165–172

Berger UV, Hediger MA (2000) Distribution of the glutamate transporters GLAST and GLT-1 in rat circumventricular organs, meninges, and dorsal root ganglia. J Comp Neurol 421:385–399

Bernardini N, De Stefano ME, Tata AM, Biagioni S, Augusti-Tocco G (1998) Neuronal and non-neuronal cell populations of the avian dorsal root ganglia express muscarinic acetylcholine receptors. Int J Dev Neurosci 16:365–377

Bernardini N, Levey AI, Augusti-Tocco G (1999) Rat dorsal root ganglia express m1-m4 muscarinic receptor proteins. J Peripher Nerv Syst 4:222–232

Bertrand I, Guillain J (1933) La microglie et l'oligodendroglie ganglionnaires. C R Soc Biol 113:382–383

Bidder FH (1847) Zur Lehre von dem Verhältnis der Ganglienkörper zu den Nervenfasern. Neue Beiträge, nebst einem Anlange von A W Volkmann. Breitkopf und Haertel, Leipzig

Bombardi C, Grandis A, Nenzi A, Giurisato M, Cozzi B (2010) Immunohistochemical localization of substance P and cholecystochinin in the dorsal root ganglia and spinal cord of the bottlenose dolphin (*Tursiops truncatus*). Anat Rec 293:477–484

Braun N, Sévigny J, Robson SC, Hammer K, Hanani M, Zimmermann H (2004) Association of the Ecto-ATPase NTPDase2 with glial cells of the peripheral nervous system. Glia 45:124–132

Brizzee KR (1949) Histogenesis of the supporting tissue in the spinal and the sympathetic trunk ganglia in the chick. J Comp Neurol 91:129–146

Brown DA, Galvan M (1977) Influence of neuroglial transport on the action of γ-aminobutyric acid on mammalian ganglion cells. Br J Pharmacol 59:373–378

Buehler A (1897) Untersuchungen über den Bau der Nervenzellen. Verh Phys Ges 31:285–392

Bunge MB, Bunge RP, Peterson ER, Murray MR (1967) Light and electron microscope study of long term organized cultures of rat dorsal root ganglia. J Cell Biol 32:439–466

Burdyga G, Lal S, Spiller D, Jiang W, Thompson D, Attwood S, Saeed S, Grundy D, Varro A, Dimaline R, Dockray J (2003) Localization of orexin-1 receptors to vagal afferent neurons in the rat and humans. Gastroenterology 124:129–139

Cajal SR see Ramón y Cajal S

Campana WM, Myers RR (2003) Exogenous erythropoietin protects against dorsal root ganglion apoptosis and pain following peripheral nerve injury. Eur J Neurosci 18:1497–1506

Carlton SM, Hargett GL (2007) Colocalization of metabotropic glutamate receptors in rat dorsal root ganglion cells. J Comp Neurol 501:780–789

Carozzi VA, Canta A, Oggioni N, Ceresa C, Marmiroli P, Konvalinka J, Zoia C, Bossi M, Ferrarese C, Tredici G, Cavaletti G (2008) Expression and distribution of 'high affinity' glutamate transporters GLT1, GLAST, EAAC1 and of GCPII in the rat peripheral nervous system. J Anat 213:539–546

Carr VM, Simpson SB Jr (1978) Proliferative and degenerative events in the early development of chick dorsal root ganglia. I. Normal development. J Comp Neurol 182:727–740

Castillo C, Norcini M, Martin Hernandez LA, Correa G, Blanck TJJ, Recio-Pinto E (2013) Satellite glia cells in dorsal root ganglia express functional NMDA receptors. Neuroscience 240:135–146

Cecchini T, Ferri P, Ciaroni S, Cuppini R, Ambrogini P, Papa S, Del Grande P (1999) Postnatal proliferation of DRG non-neuronal cells in vitamin E-deficient rats. Anat Rec 256:109–115

Cece R, Petruccioli MG, Pizzini G, Cavaletti G, Tredici G (1995) Ultrastructural aspects of DRG satellite cell involvement in experimental cisplatin neuronopathy. J Submicrosc Cytol Pathol 27:417–425

Cervós-Navarro J (1960) Elektronenmikroskopische Untersuchungen an Spinalganglien. II. Satellitenzellen. Arch Psychiatr Nervenkrank 200:267–283

Cervós-Navarro J (1962) Elektronenmikroskopische Befunde an Spinalganglienzellen der Ratte nach Ischiadikotomie. IV. Int Kongr Neuropathol 2:99–104. Thieme, Stuttgart

Chang LW, Hartmann HA (1972) Ultrastructural studies of the nervous system after mercury intoxication. I. Pathological changes in the nerve cell bodies. Acta Neuropathol 20:122–138

Chang PL, Taylor JJ, Wozniak W, Young PA (1973) An ultrastructural study of sympathetic ganglion satellite cells in the rat. I. Normal and X-ray irradiated satellite cells. J Neural Transm 34:215–234

Cherkas PS, Huang T-Y, Pannicke T, Tal M, Reichenbach A, Hanani M (2004) The effects of axotomy on neurons and satellite glial cells in mouse trigeminal ganglion. Pain 110: 290–298

Christie K, Koshy D, Cheng C, Guo G, Martinez JA, Duraikannu A, Zochodne DW (2015) Intraganglionic interactions between satellite cells and adult sensory neurons. Mol Cell Neurosci 67:1–12

Chudler EH, Anderson LC, Byers MR (1997) Trigeminal ganglion neuronal activity and glial fibrillary acidic protein immunoreactivity after inferior alveolar nerve crush in the adult rat. Pain 73:141–149

Ciaroni S, Cecchini T, Cuppini R, Ferri P, Ambrogini P, Bruno C, Del Grande P (2000) Are there proliferating neuronal precursors in adult rat dorsal root ganglia? Neurosci Lett 281:69–71

Citkowitz E, Holtzman E (1973) Peroxisomes in dorsal root ganglia. J Histochem Cytochem 21:34–41

Coggeshall RE (1967) A light and electron microscope study of the abdominal ganglion of Aplysia californica. J Neurophysiol 30:1263–1287

Cook ML, Stevens JG (1973) Pathogenesis of herpetic neuritis and ganglionitis in mice: evidence for intra-axonal transport of infection. Infect Immun 7:272–288

Cooper E (1984) Synapse formation among developing sensory neurones from rat nodose ganglia grown in tissue culture. J Physiol 351:263–274

Copray JCVM, Mantingh I, Brouwer N, Biber K, Küst BM, Liem RSB, Huitinga I, Tilders FJH, Van Dam A-M, Boddeke HWGM (2001) Expression of interleukin-1 beta in rat dorsal root ganglia. J Neuroimmunol 118:203–211

Corsetti G, Rodella L, Rezzani R, Stacchiotti A, Bianchi R (2000) Cytoplasmic changes in satellite cells of spinal ganglia induced by cisplatin treatment in rats. Ultrastruct Pathol 24:259–265

Courvoisier LG (1868) Über die Zellen der Spinalganglien sowie des Sympathicus beim Frosch. Arch Mikr Anat 4:125–145

Cravioto H, Merker HJ (1963) Elektronenmikroskopische Untersuchungen an Satellitenzellen der sympathischen Ganglien des Menschen. Arch Psychiatr Nervenkr 204:1–10

D'Amico-Martel A, Noden DM (1983) Contributions of placodal and neural crest cells to avian cranial peripheral ganglia. Am J Anat 166:445–468

De Castro F (1921) Estudio sobre los ganglios sensitivos del hombre en estado normal y patológico. Formas celulares típicas y atípicas. Trab Lab Invest Biol Univ Madrid 19:241–340

De Castro F (1932) Sensory ganglia of the cranial and spinal nerves. Normal and pathological. In: Penfield W (ed) Cytology and cellular pathology of the nervous system, vol 1. Hoeber, New York, pp 91–143

De Castro F (1946) Sobre el comportamiento y significacion de la oligodendroglia en la substancia gris central y de los gliocitos en los ganglios nerviosos perifericos. Arch Histol (Buenos Aires) 2:317–343

de Groat WC (1972) GABA-depolarization of a sensory ganglion: antagonism by picrotoxin and bicuculline. Brain Res 38:429–432 de Groat WC, Lalley PM, Saum WR (1972) Depolarization of dorsal root ganglia in the cat by GABA and related amino acids: antagonism by picrotoxin and bicuculline. Brain Res 44:273–277

De Koninck P, Carbonetto S, Cooper E (1993) NGF induces neonatal rat sensory neurons to extend dendrites in culture after removal of satellite cells. J Neurosci 13:577–585

Della Pietra V (1937) Contributo allo studio istologico dei gangli spinali. Riv Neurol 10:588–595

Detwiler SR (1937) Application of vital dyes to the study of sheath cell origin. Proc Soc Exp Biol Med 37:380–382

Devor M, Seltzer Z (1999) Pathophysiology of damaged nerves in relation to chronic pain. In: Wall PD, Melzack R (eds) Textbook of pain, 4th edn. Churcill Livingston, London, pp 129–164

Devor M, Govrin-Lippmann R, Frank I, Raber P (1985) Proliferation of primary sensory neurons in adult rat dorsal root ganglion and the kinetics of retrograde cell loss after sciatic nerve section. Somatosens Res 3:139–167

Dillard SH, Cheatham WJ, Moses HL (1972) Electron microscopy of zosteriform herpes simplex infection in the mouse. Lab Invest 26:391–402

Dixon JS (1966) The fine structure of parasympathetic nerve cells in the otic ganglia of the rabbit. Anat Rec 156:239–252

Djukic B, Casper KB, Philpot BD, Chin L-S, McCarthy KD (2007) Conditional knock-out of $K_{ir}4.1$ leads to glial membrane depolarization, inhibition of potassium and glutamate uptake, and enhanced short-term synaptic potentiation. J Neurosci 27:11354–11365

Dogiel AS (1896) Der Bau der Spinalganglien bei den Säugetieren. Anat Anz 12:140–152

Dogiel AS (1897) Zur Frage über den feineren Bau der Spinalganglien und deren Zellen bei Säugetieren. Int Monatsschr Anat Physiol 14:73–116

Dohrn A (1891) Nervenfaser und Ganglienzelle. Histogenetische Untersuchungen. Mittheil Zool Station Neapel 10:255–341

Donegan M, Kernisant M, Cua C, Jasmin L, Ohara PT (2013) Satellite glial cell proliferation in the trigeminal ganglia after chronic constriction injury of the infraorbital nerve. Glia 61:2000–2008

Donelli G, D'Uva V, Paoletti L (1975) Ultrastructure of gliosomes in ependymal cells of the lizard. J Ultrastruct Res 50:253–263

Dublin P, Hanani M (2007) Satellite glial cells in sensory ganglia: their possible contribution to inflammatory pain. Brain Behav Immun 21:592–598

Dubois-Dalcq M, Menu R, Buyse M (1972) Influence of fatty acids on fine structure of cultured neurons. An experimental approach to Refsum's disease. J Neuropathol Exp Neurol 31:645–667

Dubový P, Klusáková I, Svíženská I, Brázda V (2010) Satellite glial cells express IL-6 and corresponding signal-transducing receptors in the dorsal root ganglia of rat neuropathic pain model. Neuron Glia Biol 6:73–83

Duchen LW, Scaravilli F (1977) Quantitative and electron microscopic studies of sensory ganglion cells of the Sprawling mouse. J Neurocytol 6:465–481

Eames RA, Gamble HJ (1970) Schwann cell relationships in normal human cutaneous nerves. J Anat 106:417–435

Ebendal T (1975) Effects of nerve growth factor on the synthesis of nucleic acids and proteins in cultured chick embryo trigeminal ganglia. Zoon 3:159–167

Ehrlich P (1886) Ueber die Methylenblaureaction der lebenden Nervensubstanz. Dtsch Med Wochenschr 12:49–52

Elfvin L-G, Forsman C (1978) The ultrastructure of junctions between satellite cells in mammalian sympathetic ganglia as revealed by freeze-etching. J Ultrastruct Res 63:261–274

Elfvin L-G, Björklund H, Dahl D, Seiger A (1987) Neurofilament-like and glial fibrillary acidic protein-like immunoreactivities in rat and guinea-pig sympathetic ganglia in situ and after perturbation. Cell Tissue Res 250:79–86

Elson K, Speck P, Simmons A (2003) Herpes simplex virus infection of murine sensory ganglia induces proliferation of neuronal satellite cells. J Gen Virol 84:1079–1084

Elson K, Ribeiro RM, Perelson AS, Simmons A, Speck P (2004) The life span of ganglionic glia in murine sensory ganglia estimated by uptake of bromodeoxyuridine. Exp Neurol 186:99–103

England S, Heblich F, James IF, Robbins J, Docherty RJ (2001) Bradykinin evokes a Ca^{2+}-activated chloride current in non-neuronal cells isolated from neonatal rat dorsal root ganglia. J Physiol (London) 530:395–403

Esiri MM, Reading MC (1989) Macrophages, lymphocytes and major histocompatibility complex (HLA) class II antigens in adult human sensory and sympathetic ganglia. J Neuroimmunol 23:187–193

Farquhar MG, Hartmann JF (1957) Neuroglial structure and relationships as revealed by electron microscopy. J Neuropathol Exp Neurol 16:18–39

Feldman-Goriachnik R, Hanani M (2011) Functional study of endothelin B receptors in satellite glial cells in trigeminal ganglia. Neuroreport 22:465–469

Feldman-Goriachnik R, Belzer V, Hanani M (2015) Systemic inflammation activates satellite glial cells in the mouse nodose ganglion and alters their functions. Glia 63:2121–2132

Feltz P, Rasminsky M (1974) A model for the mode of action of GABA on primary afferent terminals: depolarizing effects of GABA applied iontophoretically to neurones of mammalian dorsal root ganglia. Neuropharmacology 13:553–563

Fenzi F, Benedetti MD, Moretto G, Rizzuto N (2001) Glial cell and macrophage reactions in rat spinal ganglion after peripheral nerve lesions: an immunocytochemical and morphometric study. Arch Ital Biol 139:357–365

Fieandt H von (1910) Eine neue Methode zur Darstellung des Gliagewebes, nebst Beiträgen zur Kenntnis des Baues und der Anordnung der Neuroglia des Hundehirns. Arch Mikr Anat 76:125–209

Field HJ, Hill TJ (1974) The pathogenesis of pseudorabies in mice following peripheral inoculation. J Gen Virol 23:145–157

Fildes Brosnan C, Bunge MB, Murray MR (1970) The response of lysosomes in cultured neurons to chlorpromazine. J Neuropathol Exp Neurol 29:337–353

Fleischhauer K (1958) Über die Feinstruktur der Faserglia. Z Zellforsch 47:548–556

Flemming W (1895) Über den Bau der Spinalganglienzellen bei Säugethieren, und Bemerkungen über den der centralen Zellen. Arch Mikr Anat 46:379–394

Forssmann WG (1964) Studien über den Feinbau des Ganglion cervicale superius der Ratte. Acta Anat 59:106–140

Forssmann WG, Tinguely H, Posternak JM, Rouiller C (1966) L'ultrastructure du ganglion cervical supérieur du rat. Les effets des rayons X. Z Zellforsch 72:325–343

Fraentzel O (1867) Beitrag zur Kenntniss von der Structur der spinalen und sympathischen Ganglienzellen. Virchows Arch 38:549–558

Freeman SE, Patil VV, Durham PL (2008) Nitric oxide-proton stimulation of trigeminal ganglion neurons increases mitogen-activated protein kinase and phosphatase expression in neurons and satellite glial cells. Neuroscience 157:542–555

Friede RL, Johnstone MA (1967) Responses of thymidine labeling of nuclei in gray matter and nerve following sciatic transection. Acta Neuropathol 7:218–231

Gabella G, Trigg P, McPhail H (1988) Quantitative cytology of ganglion neurons and satellite glial cells in the superior cervical ganglion of the sheep. Relationship with ganglion neuron size. J Neurocytol 17:753–769

Gaik GC, Farbman AI (1973) The chicken trigeminal ganglion. II. Fine structure of the neurons during development. J Morphol 141:57–75

Gallego R, Eyzaguirre C (1978) Membrane and action potential characteristics of A and C nodose ganglion cells studied in whole ganglia and in tissue slices. J Neurophysiol 41:1217–1232

Gehrmann J, Monaco S, Kreutzberg GW (1991) Spinal cord microglial cells and DRG satellite cells rapidly respond to transection of the rat sciatic nerve. Restor Neurol Neurosci 2: 181–198

Gill JS, Windebank AJ (1998) Paracrine production of nerve growth factor during rat dorsal root ganglion development. Neurosci Lett 251:149–152

Glees P, Gopinath G (1973) Age changes in the centrally and peripherally located sensory neurons in rat. Z Zellforsch 141:285–298

Glover RA (1982) Chronological changes in acid phosphatase activity within neurons and perineuronal satellite cells of the inferior vagal ganglion of the cat induced by vagotomy. J Anat 134:215–225

Gonçalves NP, Costelha S, Saraiva MJ (2014) Glial cells in familial amyloidotic polyneuropathy. Acta Neuropathol Commun 2:177

Gotow T, Yoshikawa H, Hashimoto PH (1985) Distribution patterns of orthogonal arrays and alkaline phosphatase in plasma membranes of satellite cells in rat spinal ganglia. Anat Embryol 171:171–179

Graus F, Campo E, Cruz-Sanchez F, Ribalta T, Palacin A (1990) Expression of lymphocyte, macrophage and class I and II major histocompatibility complex antigens in normal human dorsal root ganglia. J Neurol Sci 98:203–211

Gray EG (1959) Electron microscopy of neuroglial fibrils of the cerebral cortex. J Biophys Biochem Cytol 6:121–122

Gray EG (1960) Regular organisation of material in certain mitochondria of neuroglia of lizard brain. J Biophys Biochem Cytol 8:282–285

Grillo MA, Palay SL (1963) Ciliated Schwann cells in the autonomic nervous system of the adult rat. J Cell Biol 16:430–436

Grode ML, Murray MR (1973) Effects of methadone-HCl on dorsal root ganglia in organotypic culture. Exp Neurol 40:68–81

Groneberg DA, Döring F, Nickolaus M, Daniel H, Fischer A (2001) Expression of PEPT2 peptide transporter mRNA and protein in glial cells of rat dorsal root ganglia. Neurosci Lett 304:181–184

Grothe C, Meisinger C, Hertenstein A, Kurz H, Wewetzer K (1997) Expression of fibroblast growth factor-2 and fibroblast growth factor receptor 1 messenger RNAs in spinal ganglia and sciatic nerve: regulation after peripheral nerve lesion. Neuroscience 76:123–135

Gu Y, Chen Y, Zhang X, Li G-W, Wang C, Huang L-YM (2010) Neuronal soma-satellite glial cell interactions in sensory ganglia and the participation of purinergic receptors. Neuron Glia Biol 6:53–62

Hagedorn L, Paratore C, Brugnoli G, Baert J-L, Mercader N, Suter U, Sommer L (2000) The Ets domain transcription factor Erm distinguishes rat satellite glia from Schwann cells and is regulated in satellite cells by neuregulin signaling. Dev Biol 219:44–58

Hamburger V (1961) Experimental analysis of the dual origin of the trigeminal ganglion in the chick embryo. J Exp Zool 148:91–123

Hammarberg H, Piehl F, Cullheim S, Fjell J, Hökfelt T, Fried K (1996) GDNF mRNA in Schwann cells and DRG satellite cells after chronic sciatic nerve injury. Neuroreport 7:857–860

Hanani M (2005) Satellite glial cells in sensory ganglia: from form to function. Brain Res Rev 48:457–476

Hanani M, Huang TY, Cherkas PS, Ledda M, Pannese E (2002) Glial cell plasticity in sensory ganglia induced by nerve damage. Neuroscience 114:279–283

Hanker JS, Romanovicz DK, Moore GH (1974) Peroxisomes in satellite, Schwann and laminar cells associated with trigeminal sensory neurons. J Cell Biol 63:131a

Hannover A (1840) Die Chromsäure, ein vorzügliches Mittel bei mikroskopischen Untersuchungen. Arch Anat Physiol Wiss Med 549–558

Hannover A (1844) Recherches microscopiques sur le système nerveux. Philipsen, Brockhaus, Avenarius, Copenhagen

Harrison RG (1904) Neue Versuche und Beobachtungen über die Entwicklung der peripheren Nerven der Wirbeltiere. Sitzungsbericht der niederrheinischen Gesellschaft für Natur und Heilkunde, Bonn, S 55–62

Harvarik R (1977) Histochemische Untersuchungen über Veränderungen der Aktivität von Dehydrogenasen und Carboxylsäureesterasen in Spinalganglienzellen und Mantelzellen der Ratte nach Durchschneidung des Nervus ischiadicus. Dissertation, Frankfurt/M

Hauw J-J, Boutry J-M, Hamam S, Escourolle R (1978) Lipidose médicamenteuse induite en culture de ganglion spinal de souris par le maléate de perhexiline. Résultats préliminaires concernant la toxicité aiguë du medicament. C R Acad Sci (D) 287:959–961

Held H (1909) Ueber die Neuroglia marginalis der menschlichen Grosshirnrinde. Monats Psych Neurol 26(Ergänzungsheft):360–416

Hendelman W (1969) The effect of thallium on peripheral nervous tissue in culture: a light and electron microscopic study. Anat Rec 163:198–199

Herman SP, Klein R, Talley FA, Krigman MR (1973) An ultrastructural study of methylmercury-induced primary sensory neuropathy in the rat. Lab Invest 28:104–118

Herzog E (1954) Über die periphere Glia in den sympathischen Ganglien. Z Zellforsch 40:199–206

Hess A (1955) The fine structure of young and old spinal ganglia. Anat Rec 123:399–424

Hibino H, Horio Y, Fujita A, Inanobe A, Doi K, Gotow T, Uchiyama Y, Kubo T, Kurachi Y (1999) Expression of an inwardly rectifying K^+ channel, Kir4.1, in satellite cells of rat cochlear ganglia. Am J Physiol 277:C638–C644

Hill TJ, Field HJ (1973) The interaction of herpes simplex virus with cultures of peripheral nervous tissue: an electron microscopic study. J Gen Virol 21:123–133

Hinds JW, Ruffett TL (1971) Cell proliferation in the neural tube: an electron microscopic and Golgi analysis in the mouse cerebral vesicle. Z Zellforsch 115:226–264

Hirakawa H, Okajima S, Nagaoka T, Kubo T, Takamatsu T, Oyamada M (2004) Regional differences in blood-nerve barrier function and tight-junction protein expression within the rat dorsal root ganglion. Neuroreport 15:405–408

Holmgren E (1901) Beiträge zur Morphologie der Zelle. I. Nervenzellen. Anat Hefte 18:267–325

Holmgren E (1902) Einige Worte über das "Trophospongium" verschiedener Zellarten. Anat Anz 20:433–440

Holton B, Weston JA (1982) Analysis of glial cell differentiation in peripheral nervous tissue. I. S100 accumulation in quail embryo spinal ganglion cultures. Dev Biol 89:64–71

Holtzman E, Peterson ER (1969) Uptake of protein by mammalian neurons. J Cell Biol 40:863–869

Hösli E, Hösli L (1978) Autoradiographic localization of the uptake of [^{3}H]-GABA and [^{3}H] L-glutamic acid in neurones and glial cells of cultured dorsal root ganglia. Neurosci Lett 7:173–176

Hösli L, Andrès PF, Hösli E (1977) Action of GABA on neurones and satellite glial cells of cultured rat dorsal root ganglia. Neurosci Lett 6:79–83

Hösli L, Andrès PF, Hösli E (1978) Neuron-glia interactions: indirect effect of GABA on cultured glial cells. Exp Brain Res 33:425–434

Hösli L, Andrès PF, Hösli E (1979) Action of amino acid transmitters on cultured glial cells of the mammalian peripheral and central nervous system. J Physiol (Paris) 75:655–659

Hossack J, Wyburn GM (1954) Electron microscopic studies of spinal ganglion cells. Proc R Soc Edinb B 65:239–250

Huang T-Y, Cherkas PS, Rosenthal DW, Hanani M (2005) Dye coupling among satellite glial cells in mammalian dorsal root ganglia. Brain Res 1036:42–49

Huang T-Y, Hanani M, Ledda M, De Palo S, Pannese E (2006) Aging is associated with an increase in dye coupling and in gap junction number in satellite glial cells of murine dorsal root ganglia. Neuroscience 137:1185–1192

Huang T-Y, Belzer V, Hanani H (2010) Gap junctions in dorsal root ganglia: possible contribution to visceral pain. Eur J Pain 14:49.e1–49.e11

Huang L-YM, Gu Y, Chen Y (2013) Communication between neuronal somata and satellite glial cells in sensory ganglia. Glia 61:1571–1581

Huerta JJ, Diaz-Trelles R, Naves FJ, Llamosas MM, Del Valle ME, Vega JA (1996) Epidermal growth factor receptor in adult human dorsal root ganglia. Anat Embryol 194:253–257

Humbertson A Jr, Zimmermann E, Leedy M (1969) A chronological study of mitotic activity in satellite cell hyperplasia associated with chromatolytic neurons. Z Zellforsch 100:507–515

Jacob C (2015) Transcriptional control of neural crest specification into peripheral glia. Glia 63:1883–1896

Jacobs JM, Carmichael N, Cavanagh JB (1975) Ultrastructural changes in the dorsal root and trigeminal ganglia of rats poisoned with methyl mercury. Neuropathol Appl Neurobiol 1:1–19

Jacobs JM, MacFarlane RM, Cavanagh JB (1976) Vascular leakage in the dorsal root ganglia of the rat, studied with horseradish peroxidase. J Neurol Sci 29:95–107

Jancsó G, Kiraly E, Jancsó-Gábor A (1977) Pharmacologically-induced selective degeneration of chemosensitive primary sensory neurones. Nature 270:741–743

Janota I (1972) Ultrastructural studies of a hereditary sensory neuropathy in mice (dystonia musculorum). Brain 95:529–536

Jasmin L, Vit J-P, Bhargava A, Ohara PT (2010) Can satellite glial cells be therapeutic targets for pain control? Neuron Glia Biol 6:63–71

Jessen KR, Thorpe R, Mirsky R (1984) Molecular identity, distribution and heterogeneity of glial fibrillary acidic protein: an immunoblotting and immunohistochemical study of Schwann cells, satellite cells, enteric glia and astrocytes. J Neurocytol 13:187–200

Jimenez-Andrade JM, Peters CM, Mejia NA, Ghilardi JR, Kuskowski MA, Mantyh PW (2006) Sensory neurons and their supporting cells located in the trigeminal, thoracic and lumbar ganglia differentially express markers of injury following intravenous administration of paclitaxel in the rat. Neurosci Lett 405:62–67

Jones DS (1939) Studies on the origin of sheath cells and sympathetic ganglia in the chick. Anat Rec 73:343–357

Joó F, Szolcsányi J, Jancsó-Gábor A (1969) Mitochondrial alterations in the spinal ganglion cells of the rat accompanying the long-lasting sensory disturbance induced by capsaicin. Life Sci 8:621–626

Journey LJ, Burdman J, George P (1968) Ultrastructural studies on tissue culture cells treated with vincristine (NSC-67574). Cancer Chemother Rep 52:509–517

Julius D, Basbaum AI (2001) Molecular mechanisms of nociception. Nature 413:203–210

Kar S, Chabot J-G, Quirion R (1991) Quantitative autoradiographic localisation of [^{125}I]endothelin-1 binding sites in spinal cord and dorsal root ganglia of the rat. Neurosci Lett 133: 117–120

Kawamata T, Ninomiya T, Toriyabe M, Yamamoto J, Niiyama Y, Omote K, Namiki A (2006) Immunohistochemical analysis of acid-sensing ion channel 2 expression in rat dorsal root ganglion and effects of axotomy. Neuroscience 143:175–187

Key A, Retzius G (1873) Studien in der Anatomie des Nervensystems. Arch Mikr Anat 9:308–386

Kiya T, Kawamata T, Namiki A, Yamakage M (2011) Role of satellite cell-derived L-serine in the dorsal root ganglion in paclitaxel-induced painful peripheral neuropathy. Neuroscience 174:190–199

Kobayashi K, Fukuoka T, Yamanaka H, Dai Y, Obata K, Tokunaga A, Noguchi K (2006) Neurons and glial cells differentially express P2Y receptor mRNAs in the rat dorsal root ganglion and spinal cord. J Comp Neurol 498:443–454

Kobayashi T, Yamauchi K, Matsuura Y, Kuniyoshi K, Takahashi K, Ohtori S (2015) The effects of generally administered anti-NGF receptor (p75NTR) antibody on pain-related behavior, dorsal root ganglia, and spinal glia activation in a rat model of brachial plexus avulsion. J Hand Surg Am 40:2017–2025

Koelliker A (1905) Die Entwicklung der Elemente des Nervensystems. Z Wissen Zool 82:1–38

Koeppen AH, Morral JA, Davis AN, Qian J, Petrocine SV, Knutson MD, Gibson WM, Cusack MJ, Li D (2009) The dorsal root ganglion in Friedreich's ataxia. Acta Neuropathol 118:763–776

Kohn A (1907) Über die Scheidenzellen (Randzellen) peripherer Ganglienzellen. Anat Anz 30:154–159

Kohno K (1969) Ultrastructure of long gliosome in satellite cell of frog spinal ganglion. Bull Tokyo Med Dent Univ 16:303–309

Koike T, Wakabayashi T, Mori T, Takamori Y, Hirahara Y, Yamada H (2014) Sox2 in the adult rat sensory nervous system. Histochem Cell Biol 141:301–309

Koneff H (1887) Beiträge zur Kenntnis in den Nervenzellen der peripheren Ganglien. Mitt Naturforsch Ges Bern, Nr 1143–1168, S 13–44

Krajčí D (1973) Ontogenetic development of the relation between neurons and satellite cells in spinal ganglia. Folia Morphol (Warsz) 21:139–141

Krajčí D (1975) Unusual intracapsular and interstitial cells in spinal ganglia of cat and their interrelationships. Acta Univ Palacki Olomuc Fac Med 73:165–176

Kraus-Ruppert R, Laissue J, Bürki H, Odartchenko N (1975) Kinetic studies on glial, Schwann and capsular cells labelled with [^{3}H]thymidine in cerebrospinal tissue of young mice. J Neurol Sci 26:555–563

Krawczyk WS, Wilgram GF (1973) Hemidesmosome and desmosome morphogenesis during epidermal wound healing. J Ultrastruct Res 45:93–101

Kubota K, Hioki K (1943) Zytologische Untersuchungen der Mantelzellen im menschlichen Spinalganglion. Okajimas Folia Anat Jpn 22:111–126

Kummer W, Behrends S, Schwarzlmüller T, Fischer A, Koesling D (1996) Subunits of soluble guanylyl cyclase in rat and guinea-pig sensory ganglia. Brain Res 721:191–195

Kung L-H, Gong K, Adedoyin M, Ng J, Bhargava A, Ohara PT, Jasmin L (2013) Evidence for glutamate as a neuroglial transmitter within sensory ganglia. PLoS One 8:e68312

Kuntz A, Sulkin NM (1947) The neuroglia in the autonomic ganglia: cytologic structure and reactions to stimulation. J Comp Neurol 86:467–477

Kuo L-T, Simpson A, Schänzer A, Tse J, An S-F, Scaravilli F, Groves MJ (2005) Effects of systemically administered NT-3 on sensory neuron loss and nestin expression following axotomy. J Comp Neurol 482:320–332

Lahl R (1975) Die Pathomorphologie des ZNS bei der Tetrachlorkohlenstoffintoxication. Zentralbl Allg Pathol 119:276–285

LaVail JH, Topp KS, Giblin PA, Garner JA (1997) Factors that contribute to the transneuronal spread of herpes simplex virus. J Neurosci Res 49:485–496

Ledda M, Barni L, Altieri L, Pannese E (1999) Amount and distribution of lipofuscin in nerve and satellite cells from spinal ganglia of young adult and aged rabbits. J Submicrosc Cytol Pathol 31:237–246

Ledda M, Barni L, Altieri L, Pannese E (2003) The Golgi apparatus of satellite cells associated with spinal ganglion neurons: changes with age in the rabbit. J Submicrosc Cytol Pathol 35:267–270

Leech RW (1967) Changes in satellite cells of rat dorsal root ganglia during central chromatolysis. An electron microscopic study. Neurology 17:349–358

Lenghaus C, Mann JA, Done JT, Bradley R (1976) Neuropathology of experimental swine vesicular disease in pigs. Res Vet Sci 21:19–27

Lenhossék M von (1897) Über den Bau der Spinalganglienzellen des Menschen. Arch Psychiatr Nervenkr 29:345–380

Lenhossék M von (1907) Zur Kenntniss der Spinalganglienzellen. Arch Mikr Anat 69:245–263

Levi G (1907) La capsula delle cellule dei gangli sensitivi. Penetrazione di fibre collagene nel loro protoplasma. Monit Zool Ital 18:153–158

Levi G (1908) I gangli cerebrospinali. Arch Ital Anat Embriol 7(Suppl):1–392

Levin MJ, Cai G-Y, Manchak MD, Pizer LI (2003) Varicella-zoster virus DNA in cells isolated from human trigeminal ganglia. J Virol 77:6979–6987

Levy BDFA, Cunha JDC, Chadi G (2007) Cellular analysis of S100β and fibroblast growth factor-2 in the dorsal root ganglia and sciatic nerve of rodents. Focus on paracrine actions of activated satellite cells after axotomy. Int J Neurosci 117:1481–1503

Leydig F (1851) Zur Anatomie und Histologie der Chimaera monstrosa. Arch Anat Physiol Wiss Med 241–271

Li M, Shi J, Tang J-r, Chen D, Ai B, Chen J, Wang L-n, Cao F-y, Li L-l, Lin C-y, Guan X-m (2005) Effects of complete Freund's adjuvant on immunohistochemical distribution of IL-1β and IL-1R I in neurons and glia cells of dorsal root ganglion. Acta Pharmacol Sin 26:192–198

Li J, Vause CV, Durham PL (2008) Calcitonin gene-related peptide stimulation of nitric oxide synthesis and release from trigeminal ganglion glial cells. Brain Res 1196:22–32

Lieberman AR (1971) The axon reaction: a review of the principal features of perikaryal responses to axon injury. Int Rev Neurobiol 14:49–124

Lieberman AR (1976) Sensory ganglia. In: Landon DN (ed) The peripheral nerve. Chapman and Hall, London, pp 188–278

Lindner G, Grosse G (1974) Morphometrische Untersuchungen am Ganglion trigeminale vom Hühnerembryo in situ and in der In-vitro-Kultur. Z Mikr Anat Forsch 88:439–448

Liu W, Glueckert R, Linthicum FH, Rieger G, Blumer M, Bitsche M, Pechriggl E, Rask-Andersen H, Schrott-Fischer A (2014) Possible role of gap junction intercellular channels and connexin 43 in satellite glial cells (SGCs) for preservation of human spiral ganglion neurons. Cell Tissue Res 355:267–278

Lodin Z, Booher J, Kasten FH (1970) Phase-contrast cinematographic study of dissociated neurons from embryonic chick dorsal root ganglia cultured in the Rose chamber. Exp Cell Res 60:27–39

Lodin Z, Faltin J, Booher J, Hartman J, Sensenbrenner M (1973) Formation of intercellular contacts in cultures of dissociated neurons from embryonic chicken dorsal root ganglia. An electron microscopic and scanning electron microscopic study. Neurobiology 3:376–390

Low FN (1970) Interstitial bodies in the early chick embryo. Am J Anat 128:45–56

Lu X, Richardson PM (1991) Inflammation near the nerve cell body enhances axonal regeneration. J Neurosci 11:972–978

Magnusson S, Ekström J, Elmér E, Kanje M, Ny L, Alm P (2000) Heme oxigenase-1, heme oxigenase-2 and biliverdin reductase in peripheral ganglia from rat, expression and plasticity. Neuroscience 95:821–829

Mannu A (1935) Ricerche sulla evoluzione dei neuroni nei gangli spinali dei mammiferi (Bos taurus). Riv Biol 19:225–250

Martinelli C, Sartori P, De Palo S, Ledda M, Pannese E (2005) Increase in number of the gap junctions between satellite neuroglial cells during lifetime: an ultrastructural study in rabbit spinal ganglia from youth to extremely advanced age. Brain Res Bull 67:19–23

Martinelli C, Sartori P, De Palo S, Ledda M, Pannese E (2006a) The perineuronal glial tissue of spinal ganglia. Quantitative changes in the rabbit from youth to extremely advanced age. Anat Embryol 211:455–463

Martinelli C, Sartori P, Ledda M, Pannese E (2006b) A study of mitochondria in spinal ganglion neurons during life: quantitative changes from youth to extremely advanced age. Tissue Cell 38:93–98

Martinelli C, Sartori P, Ledda M, Pannese E (2007) Mitochondria in perineuronal satellite cell sheaths of rabbit spinal ganglia: quantitative changes during life. Cells Tissues Organs 186:141–146

Masaki T, Matsumura K, Hirata A, Yamada H, Hase A, Shimizu T, Yorifuji H, Motoyoshi K, Kamakura K (2001) Expression of dystroglycan complex in satellite cells of dorsal root ganglia. Acta Neuropathol 101:174–178

Masurovsky EB, Bunge MB, Bunge RP (1967) Cytological studies of organotypic cultures of rat dorsal root ganglia following X-irradiation in vitro. I. Changes in neurons and satellite cells. J Cell Biol 32:467–496

Masurovsky EB, Peterson ER, Crain SM, Horwitz SB (1983) Morphological alterations in dorsal root ganglion neurons and supporting cells of organotypic mouse spinal cord-ganglion cultures exposed to taxol. Neuroscience 10:491–509

Matsuda S, Kobayashi N, Terashita T, Shimokawa T, Shigemoto K, Mominoki K, Wakisaka H, Saito S, Miyawaki K, Saito K, Kushihata F, Chen J, Gao S-Y, Li C-Y, Wang M, Fujiwara T (2005) Phylogenetic investigation of Dogiel's pericellular nests and Cajal's initial glomeruli in the dorsal root ganglion. J Comp Neurol 491:234–245

Matsumoto E, Rosenbluth J (1986) Structure of the satellite cell sheath around the cell body, axon hillock, and initial segment of frog dorsal root ganglion cells. Anat Rec 215:182–191

Matthews MR, Raisman G (1972) A light and electron microscopic study of the cellular response to axonal injury in the superior cervical ganglion of the rat. Proc R Soc Lond B Biol Sci 181:43–79

McCracken RM, Dow C (1973a) An electron microscopic study of normal bovine spinal ganglia and nerves. Acta Neuropathol 25:127–137

McCracken RM, Dow C (1973b) An electron microscopic study of Aujeszky's disease. Acta Neuropathol 25:207–219

McLachlan EM, Jänig W, Devor M, Michaelis M (1993) Peripheral nerve injury triggers noradrenergic sprouting within dorsal root ganglia. Nature 363:543–546

Meier C, Glees P (1971) Der Einfluss des Centrophenoxins auf das Alterspigment in Satellitenzellen und Neuronen der Spinalganglien seniler Ratten. Eine elektronenmikroskopische Untersuchung. Acta Neuropathol 17:310–320

Meller K, Waelsch M (1975) Changes in glia-neuron relationships in cell cultures of spinal ganglia caused by puromycin. Cell Tissue Res 160:431–442

Miller R, Varon S, Kruger L, Coates PW, Orkand PM (1970) Formation of synaptic contacts on dissociated chick embryo sensory ganglion cells in vitro. Brain Res 24:356–358

Miller KE, Richards BA, Kriebel RM (2002) Glutamine-, glutamine synthetase-, glutamate dehydrogenase- and pyruvate carboxylase-immunoreactivities in the rat dorsal root ganglion and peripheral nerve. Brain Res 945:202–211

Mirsky R, Jessen JR, Schachner M, Goridis C (1986) Distribution of the adhesion molecules N-CAM and L1 on peripheral neurons and glia in adult rats. J Neurocytol 15:799–815

Morris R, Southam E, Braid DJ, Garthwaite J (1992) Nitric oxide may act as a messenger between dorsal root ganglion neurones and their satellite cells. Neurosci Lett 137:29–32

Moses HL, Beaver DL, Ganote CE (1965) Electron microscopy of the trigeminal ganglion. I. Comparative ultrastructure. Arch Pathol 79:541–556

Mudge AW (1981) Effect of non-neuronal cells on peptide content of cultured sensory neurones. J Exp Biol 95:195–203

Mudge AW (1984) Schwann cells induce morphological transformation of sensory neurones in vitro. Nature 309:367–369

Muratori L, Ronchi G, Raimondo S, Geuna S, Giacobini-Robecchi MG, Fornaro M (2015) Generation of new neurons in dorsal root ganglia in adult rats after peripheral nerve crush injury. Neural Plast 2015:860546

Murray MR, Peterson ER (1964) Actions of drugs and toxic substances on nervous tissue in culture. In: Richter D (ed) Comparative neurochemistry. Pergamon, Oxford, London, New York, Paris, pp 451–458

Naciff JM, Kaetzel MA, Behbehani MM, Dedman JR (1996) Differential expression of annexins I-VI in the rat dorsal root ganglia and spinal cord. J Comp Neurol 368:356–370

Nageotte J (1907a) Neurophagie dans les greffes de ganglions rachidiens. Rev Neurol 15:933–944

Nageotte J (1907b) Troisième note sur la greffe des ganglions rachidiens; mode de destruction des cellules nerveuses mortes. C R Soc Biol (Paris) 62:381–384

Nakamura Y, Iga K, Shibata T, Shudo M, Kataoka K (1993) Glial plasmalemmal vesicles: a subcellular fraction from rat hippocampal homogenate distinct from synaptosomes. Glia 9:48–56

Narayanan CH, Narayanan Y (1980) Neural crest and placodal contribution in the development of the glossopharyngeal-vagal complex in the chick. Anat Rec 196:71–82

Nathaniel EJH, Nathaniel DR (1973) Electron microscopic studies of spinal ganglion cells following crushing of dorsal roots in adult rat. J Ultrastruct Res 45:168–182

Nawzatzky I (1933) Zur Kenntnis der Farbspeicherung in peripherischen Ganglien der Maus. Z Zellforsch 20:229–236

Nemiloff A (1908) Beobachtungen über die Nervenelemente bei Ganoïden und Knochenfischen. I. Der Bau der Nervenzellen. Arch Mikr Anat 72:1–46

Newcomb EH, Steer MW, Hepler PK, Wergin WP (1968) An atypical crista resembling a "tight junction" in bean root mitochondria. J Cell Biol 39:35–42

Novikoff AB, Quintana N, Villaverde H, Forschirm R (1966) Nucleoside phosphatase and cholinesterase activities in dorsal root ganglia and peripheral nerve. J Cell Biol 29:525–545

Obersteiner EJ, Sharma RP (1978) Effect of vitamin E on selenium cytotoxicity in chick ganglia cultures. Toxicology 9:165–172

Ohara PT, Vit J-P, Bhargava A, Jasmin L (2008) Evidence for a role of connexin43 in trigeminal pain using RNA interference in vivo. J Neurophysiol 100:3064–3073

Ohara PT, Vit J-P, Bhargava A, Romero M, Sundberg C, Charles AC, Jasmin L (2009) Gliopathic pain: when satellite glial cells go bad. Neuroscientist 15:450–463

Ohtori S, Takahashi K, Moriya H, Myers RR (2004) TNF-α and TNF-α receptor type 1 upregulation in glia and neurons after peripheral nerve injury. Studies in murine DRG and spinal cord. Spine 29:1082–1088

Olsson Y (1971) Studies on vascular permeability in peripheral nerves. IV. Distribution of intravenously injected protein tracers in the peripheral nervous system of various species. Acta Neuropathol 17:114–126

Ortiz-Picón JM (1932) La oligodendroglía de los ganglios sensitívos. Rev Españ Biol 1:19–24

Ortiz-Picón JM (1949) Nouvelle contribution à l'étude de la névroglie (oligodendroglie) des ganglions sensitifs. Bull Histol Appl 26:113–123

Ortiz-Picón JM (1955) The neuroglia of the sensory ganglia. Anat Rec 121:513–529

Palay SL (1957) Contributions of electron microscopy to neuroanatomy. In: Windle WF (ed) New research techniques of neuroanatomy. Thomas, Springfield, pp 5–16

Palumbi G (1944) Osservazioni sulle capsule pericellulari e sulle cellule satelliti dei gangli spinali e simpatici. Ricerche Morf 20/21:117–144

Pannese E (1960) Observations on the morphology, submicroscopic structure and biological properties of satellite cells (s.c.) in sensory ganglia of mammals. Z Zellforsch 52:567–597

Pannese E (1963) Investigations on the ultrastructural changes of the spinal ganglion neurons in the course of axon regeneration and cell hypertrophy. II. Changes during cell hypertrophy and comparison between the ultrastructure of nerve cells of the same type under different functional conditions. Z Zellforsch 61:561–586

Pannese E (1964) Number and structure of perisomatic satellite cells of spinal ganglia under normal conditions or during axon regeneration and neuronal hypertrophy. Z Zellforsch 63:568–592

Pannese E (1968a) Developmental changes of the endoplasmic reticulum and ribosomes in nerve cells of the spinal ganglia of the domestic fowl. J Comp Neurol 132:331–364

Pannese E (1968b) Temporary junctions between neuroblasts in the developing spinal ganglia of the domestic fowl. J Ultrastruct Res 21:233–250

Pannese E (1969) Electron microscopical study on the development of the satellite cell sheath in spinal ganglia. J Comp Neurol 135:381–422

Pannese E (1974) The histogenesis of the spinal ganglia. Adv Anat Embryol Cell Biol 47(5):1–97

Pannese E (1978) The response of the satellite and other non-neuronal cells to the degeneration of neuroblasts in chick embryo spinal ganglia. Cell Tissue Res 190:1–14

Pannese E (1981) The satellite cells of the sensory ganglia. Adv Anat Embryol Cell Biol 65:1–111

Pannese E (2015) Neurocytology. Fine structure of neurons, nerve processes, and neuroglial cells, 2nd edn. Springer, New York

Pannese E, Procacci P (2002) Ultrastructural localization of NGF receptors in satellite cells of the rat spinal ganglia. J Neurocytol 31:755–763

Pannese E, Bianchi R, Calligaris B, Ventura R, Weibel ER (1972) Quantitative relationships between nerve and satellite cells in spinal ganglia. An electron microscopical study. I. Mammals. Brain Res 46:215–234

Pannese E, Ventura R, Bianchi R (1975) Quantitative relationships between nerve and satellite cells in spinal ganglia: an electron microscopical study. II. Reptiles. J Comp Neurol 160:463–476

Pannese E, Luciano L, Iurato S, Reale E (1977) Intercellular junctions and other membrane specializations in developing spinal ganglia: a freeze-fracture study. J Ultrastruct Res 60:169–180

Pannese E, Luciano L, Reale E (1978) Intercellular junctions in developing spinal ganglia. Zoon 6:129–138

Pannese E, Rigamonti L, Procacci P, Ledda M, Arcidiacono G, Frattola D (1987) An electron microscope study of quantitative relationships between axon and Schwann cell sheath in myelinated fibres of peripheral nerves. Anat Embryol 175:423–430

Pannese E, Arcidiacono G, Frattola D, Rigamonti L, Procacci P, Ledda M (1988) Quantitative relationships between axoplasm and Schwann cell sheath in unmyelinated nerve fibres. An electron microscope study. J Anat 159:49–56

Pannese E, Ledda M, Arcidiacono G, Rigamonti L (1991) Clusters of nerve cell bodies enclosed within a common connective tissue envelope in the spinal ganglia of the lizard and rat. Cell Tissue Res 264:209–214

Pannese E, Procacci P, Ledda M, Conte V (1993) The percentage of nerve cell bodies arranged in clusters decreases with age in the spinal ganglia of adult rabbits. Anat Embryol 187: 331–334

Pannese E, Rigamonti L, Ledda M, Arcidiacono G (1994) Perikaryal projections of spinal ganglion neurons: quantitative differences between membrane domains in contact with different microenvironments. J Anat 185:497–502

Pannese E, Ledda M, Conte V, Rigamonti L, Procacci P (1995) On the influence of the perineuronal microenvironment on the outgrowth of perikaryal projections of spinal ganglion neurons. J Submicrosc Cytol Pathol 27:303–308

Pannese E, Procacci P, Ledda M, Conte V (1996) Age-related reduction of the satellite cell sheath around spinal ganglion neurons in the rabbit. J Neurocytol 25:137–146

Pannese E, Ledda M, Martinelli C, Sartori P (1997) Age-related decrease of the perineuronal satellite cell number in the rabbit spinal ganglia. J Peripher Nerv Syst 2:77–82

Pannese E, Procacci P, Berti E, Ledda M (1999) The perikaryal surface of spinal ganglion neurons: differences between domains in contact with satellite cells and in contact with the extracellular matrix. Anat Embryol 199:199–206

Pannese E, Ledda M, Cherkas PS, Huang TY, Hanani M (2003) Satellite cell reactions to axon injury of sensory ganglion neurons. Increase in number of gap junctions and formation of bridges connecting previously separate perineuronal sheaths. Anat Embryol 206:337–347

Patterson PH, Chun LLY (1974) The influence of non-neuronal cells on catecholamine and acetylcholine synthesis and accumulation in cultures of dissociated sympathetic neurons. Proc Natl Acad Sci U S A 71:3607–3610

Pazour GJ, Witman GB (2003) The vertebrate primary cilium is a sensory organelle. Curr Opin Cell Biol 15:105–110

Penfield W (1932) Tumors of the sheaths of the nervous system. In: Penfield W (ed) Cytology and cellular pathology of the nervous system, vol 3. Hoeber, New York, pp 953–990

Penta P (1934) Osservazioni sulla capsula nei gangli spinali. Riv Patol Nerv Ment 44:509–513

Pérez Armendariz E M, Norcini M, Hernández-Tellez B, Castell-Rodríguez A, Coronel-Cruz C, Guerrero Alquicira R, Sideris A, Recio-Pinto E (2018) Neurons and satellite glial cells in adult rat lumbar dorsal root ganglia express connexin 36. Acta Histochem 120:168–178

Peters CM, Jimenez-Andrade JM, Jonas BM, Sevcik MA, Koewler NJ, Ghilardi JR, Wong GY, Mantyh PW (2007) Intravenous paclitaxel administration in the rat induces a peripheral sensory neuropathy characterized by macrophage infiltration and injury to sensory neurons and their supporting cells. Exp Neurol 203:42–54

Pick J (1963) The submicroscopic organization of the sympathetic ganglion in the frog (Rana pipiens). J Comp Neurol 120:409–462

Pilar G, Landmesser L (1976) Ultrastructural differences during embryonic cell death in normal and peripherally deprived ciliary ganglia. J Cell Biol 68:339–356

Pineda A, Maxwell DS, Kruger L (1967) The fine structure of neurons and satellite cells in the trigeminal ganglion of cat and monkey. Am J Anat 121:461–488

Pomerat CM, Hendelman WJ, Raiborn CW Jr, Massey JF (1967) Dynamic activities of nervous tissue in vitro. In: Hydén H (ed) The neuron. Elsevier, Amsterdam, pp 119–178

Pomeroy SL, Zurakowski D, Khoxayo S, Endara M, Dikkes P (1996) Postnatal addition of satellite cells to parasympathetic neurons. J Comp Neurol 375:518–525

Pomonis JD, Rogers SD, Peters CM, Ghilardi JR, Mantyh PW (2001) Expression and localization of endothelin receptors: implications for the involvement of peripheral glia in pain. J Neurosci 21:999–1006

Popken GJ, Farel PB (1997) Sensory neuron number in neonatal and adult rats estimated by means of stereologic and profile-based methods. J Comp Neurol 386:8–15

Price TJ, Hargreaves KM, Cervero F (2006) Protein expression and mRNA cellular distribution of the NKCC1 cotransporter in the dorsal root and trigeminal ganglia of the rat. Brain Res 1112:146–158

Prineas J, Spencer PS (1975) Pathology of the nerve cell body in disorders of the peripheral nervous system. In: Dyck PJ, Thomas PK, Lambert EH (eds) Peripheral neuropathy, vol 1. Saunders, Philadelphia, London, Toronto, pp 253–295

Procacci P, Magnaghi V, Pannese E (2008) Perineuronal satellite cells in mouse spinal ganglia express the gap junction protein connexin43 throughout life with decline in old age. Brain Res Bull 75:562–569

Pruginin-Bluger M, Shelton DL, Kalcheim C (1997) A paracrine effect for neuron-derived BDNF in development of dorsal root ganglia: stimulation of Schwann cell myelin protein expression by glial cells. Mech Dev 61:99–111

Quade RH (1939) A new staining technic originated to demonstrate the capsular cells of the sympathetic nervous system. Proc Mayo Clin 14:555–560

Ramón y Cajal S (1890) Sobre la existencia de terminaciones nerviosas pericelulares en los ganglios nerviosos raquidianos. Pequeñas Comun Anat (Barcelona) pp 1–5

Ramón y Cajal S (1907) Die Struktur der sensiblen Ganglien des Menschen und der Tiere. Ergebn Anat Entwickl Gesch 16:177–215

Ramón y Cajal S (1909) Histologie du système nerveux de l'homme et des vertébrés, vol 1. Maloine, Paris

Ramón y Cajal S (1912) El aparato endocelular de Golgi de la célula de Schwann y algunas observaciones sobre la estructura de los tubos nerviosos. Trab Lab Invest Biol Univ Madrid 10:221–246

Ramón y Cajal S, Oloriz F (1897) Los ganglios sensitivos craneales de los mamiferos. Rev Trim Micr 2:129–151

Rash JE, Yasumura T, Hudson CS, Agre P, Nielsen S (1998) Direct immunogold labeling of aquaporin-4 in square arrays of astrocyte and ependymocyte plasma membranes in rat brain and spinal cord. Proc Natl Acad Sci U S A 95:11981–11986

Rashid MH, Inoue M, Matsumoto M, Ueda H (2004) Switching of bradykinin-mediated nociception following partial sciatic nerve injury in mice. J Pharmacol Exp Ther 308:1158–1164

Reichelt M, Zerboni L, Arvin AM (2008) Mechanisms of varicella-zoster virus neuropathogenesis in human dorsal root ganglia. J Virol 82:3971–3983

Remak R (1838) Observationes anatomicae et microscopicae de systematis nervosi structura. Dissertation, Berolini

Retamal M A, Riquelme M A, Stehberg J, Alcayaga J (2017) Connexin43 hemichannels in satellite glial cells, can they influence sensory neuron activity? Front Mol Neurosci 10:374

Retzius G (1894) Zur Frage von den freien Nervenendigungen in den Spinalganglien. Biol Untersuch NF 6:59–61

Riegele L (1932) Beitrag zur Kenntnis des Scheidenplasmodiums im autonomen Nervensystem. Z Zellforsch 15:374–397

Río Hortega P del, Polak M, Prado JM (1942) Investigaciones sobre la neuroglia de los ganglios sensitivos. Arch Histol (Buenos Aires) 1:233–275

Rosenbluth J (1962a) The fine structure of neurons and satellite cells in spinal ganglia of the toad. Anat Rec 142:344

Rosenbluth J (1962b) The fine structure of acoustic ganglia in the rat. J Cell Biol 12:329–359

Rosenbluth J (1963) Contrast between osmium-fixed and permanganate-fixed toad spinal ganglia. J Cell Biol 16:143–157

Rosenbluth J, Palay SL (1961) The fine structure of nerve cell bodies and their myelin sheaths in the eighth nerve ganglion of the goldfish. J Biophys Biochem Cytol 9:853–877

Rosenbluth J, Wissig SL (1964) The distribution of exogenous ferritin in toad spinal ganglia and the mechanism of its uptake by neurons. J Cell Biol 23:307–325

Rozanski GM, Kim H, Li Q, Wong FK, Stanley EF (2012) Slow chemical transmission between dorsal root ganglion neuron somata. Eur J Neurosci 36:3314–3321

Rozanski GM, Nath AR, Adams ME, Stanley EF (2013) Low voltage-activated calcium channels gate transmitter release at the dorsal root ganglion sandwich synapse. J Physiol 591:5575–5583

Sakuma E, Wang HJ, Asai Y, Tamaki D, Amano K, Mabuchi Y, Herbert DC, Soji T (2001) Gap junctional communication between the satellite cells of rat dorsal root ganglia. Acta Anat Nippon 76:297–302

Sandelin M, Zabihi S, Liu L, Wicher G, Kozlova EN (2004) Metastasis-associated S100A4 (Mts1) protein is expressed in subpopulations of sensory and autonomic neurons and in Schwann cells of the adult rat. J Comp Neurol 473:233–243

Sato M, Austin G (1961) Intracellular potentials of mammalian dorsal root ganglion cells. J Neurophysiol 24:569–582

Schaffer J (1896) Über einen neuen Befund von Centrosomen in Ganglien- und Knorpelzellen. Sitzungsber Akad Wiss Wien 105:21–28

Scharenberg K (1952) Glia and the elements of Schwann of the human Gasserian ganglion. Trab Inst Cajal Invest Biol 44:75–94

Scharf JH (1958) Sensible Ganglien. In: von Möllendorf W, Bargmann W (Hrsg) Handbuch der mikroskopischen Anatomie des Menschen, Bd 4/3. Springer, S 14–15, 290–300

Schlaepfer WW (1969) Experimental lead neuropathy: a disease of the supporting cells in the peripheral nervous system. J Neuropathol Exp Neurol 28:401–418

Schlaepfer WW (1971) Cadmium-incidence injury in the rat spinal ganglia. J Neuropathol Exp Neurol 30:141–142

Schon F, Kelly JS (1974a) Autoradiographic localisation of [^{3}H]GABA and [^{3}H]glutamate over satellite glial cells. Brain Res 66:275–288

Schon F, Kelly JS (1974b) The characterisation of [^{3}H]GABA uptake into the satellite glial cells of rat sensory ganglia. Brain Res 66:289–300

Schramm J (1864) Neue Untersuchungen über den Bau der Spinalganglien. Med Inaug-Diss, Würzburg

Schröder JM (1970) Zur Pathogenese der Isoniazid-Neuropathie. II. Phasenkontrast-und elektronenmikroskopische Untersuchungen am Rückenmark, an Spinalganglien und Muskelspindeln. Acta Neuropathol 16:324–341

Schultze M (1871) Allgemeines über die Structurelemente des Nervensystems. In: Stricker S (Hrsg) Handbuch der Lehre von den Geweben des Menschen und der Thiere, Bd 1. Engelmann, Leipzig, S 108–136

Schwandt H-D (1976) Enzymhistochemische Untersuchungen über Veränderungen sauer Phosphatasen, der alkalischen Phosphatase, Adenosintriphosphatase und unspezifischen Esterase in Spinalganglienzellen und Mantelzellen der Ratte nach Durchschneidung des N. ischiadicus. Inaug-Diss, Frankfurt

Sharma K, Korade Z, Frank E (1995) Late-migrating neuroepithelial cells from the spinal cord differentiate into sensory ganglion cells and melanocytes. Neuron 14:143–152

Shi T-JS, Holmberg K, Xu Z-QD, Steinbusch H, de Vente J, Hökfelt T (1998) Effect of peripheral nerve injury on cGMP and nitric oxide synthase levels in rat dorsal root ganglia: time course and coexistence. Pain 78:171–180

Shimeld C, Whiteland JL, Williams NA, Easty DL, Hill TJ (1997) Cytokine production in the nervous system of mice during acute and latent infection with herpes simplex virus type 1. J Gen Virol 78:3317–3325

Shimizu Y (1965) The satellite cells in cultures of dissociated spinal ganglia. Z Zellforsch 67:185–195

Shinder V, Devor M (1994) Structural basis of neuron-to-neuron cross-excitation in dorsal root ganglia. J Neurocytol 23:515–531

Shinder V, Govrin-Lippmann R, Cohen S, Belenky M, Ilin P, Fried K, Wilkinson H, Devor M (1999) Structural basis of sympathetic-sensory coupling in rat and human dorsal root ganglia following peripheral nerve injury. J Neurocytol 28:743–761

Shoji Y, Yamaguchi-Yamada M, Yamamoto Y (2010) Glutamate- and GABA-mediated neuron-satellite cell interaction in nodose ganglia as revealed by intracellular calcium imaging. Histochem Cell Biol 134:13–22

Sjögreen B, Wiklund P, Ekström PAR (2000) Mitogen activated protein kinase inhibition by PD98059 blocks nerve growth factor stimulated axonal outgrowth from adult mouse dorsal root ganglia in vitro. Neuroscience 100:407–416

Skoglund S (1967) On the possible postnatal formation of new nerve fibres in the dorsal roots from new nerve cells in the ganglia. An autoradiographic study with H^3-thymidine in the cat. Acta Soc Med Upsalien 72:25–29

Spataro LE, Sloane LM, Milligan ED, Wieseler-Frank J, Schoeniger D, Jekich BM, Barrientos RM, Maier SF, Watkins LR (2004) Spinal gap junctions: potential involvement in pain facilitation. J Pain 5:392–405

Spencer PS, Peterson ER, Madrid RA, Raine CS (1973) Effects of thallium salts on neuronal mitochondria in organotypic cord-ganglia-muscle combination cultures. J Cell Biol 58: 79–95

Spoerri PE, Glees P (1974) The effects of dimethylaminoethyl p-chlorophenoxyacetate on spinal ganglia neurons and satellite cells in culture. Mitochondrial changes in the aging neurons. An electron microscope study. Mech Ageing Dev 3:131–155

Srebro Z (1965) The ultrastructure of gliosomes in the brains of amphibia. J Cell Biol 26:313–322

Stefansson K, Wollmann RL, Moore BW (1982) Distribution of S-100 protein outside the central nervous system. Brain Res 234:309–317

Stensaas LJ, Fidone SJ (1977) An ultrastructural study of cat petrosal ganglia: a search for autonomic ganglion cells. Brain Res 124:29–39

Stephenson JL, Byers MR (1995) GFAP immunoreactivity in trigeminal ganglion satellite cells after tooth injury in rats. Exp Neurol 131:11–22

Stewart HJS, Rougon G, Dong Z, Dean C, Jessen KR, Mirsky R (1995) TGF-βs upregulate NCAM and L1 expression in cultured Schwann cells, suppress cyclic AMP-induced expression of 04 and galactocerebroside, and are widely expressed in cells of the Schwann cell lineage in vivo. Glia 15:419–436

Stöhr Ph Jr (1928) Das peripherische Nervensystem. A. Die Anteile des cerebrospinalen Nervensystems. In: von Möllendorf W (Hrsg) Handbuch der mikroskopischen Anatomie des Menschen, Bd 4/1. Springer, Berlin, S 202–264

Stramignoni A (1953) Morfologia e struttura dei cosiddetti satelliti perineuronali dei gangli spinali, loro alterazioni cadaveriche e comportamento in alcune condizioni patologiche. Arch Sci Med 78:231–255

Streeter GL (1905) On the histogenesis of spinal ganglia in mammals. Am J Anat 4:XIII

Streeter GL (1912) The development of the nervous system. In: Keibel F, Mall FP (eds) Manual of human embryology, vol 2. Lippincott, Philadelphia, London, pp 1–156

Suadicani SO, Cherkas PS, Zuckerman J, Smith DN, Spray DC, Hanani M (2010) Bidirectional calcium signaling between satellite glial cells and neurons in cultured mouse trigeminal ganglia. Neuron Glia Biol 6:43–51

Sugimoto T, Takeyama A, Fujita M, Ichikawa H, Takano-Yamamoto T (2001) Peripheral neuroglial death induced by cisplatin administration in newborn rats. Neuroreport 12:137–140

Sulkin DF, Sulkin NM, Nushan H (1973) Fine structure of sensory ganglia during experimental scurvy. Acta Neuropathol 23:141–151

Sylvia AL, Rosenthal M (1979) Effects of age on brain oxidative metabolism in vivo. Brain Res 165:235–248

Takeda M, Tanimoto T, Kadoi J, Nasu M, Takahashi M, Kitagawa J, Matsumoto S (2007) Enhanced excitability of nociceptive trigeminal ganglion neurons by satellite glial cytokine following peripheral inflammation. Pain 129:155–166

Takeda M, Takahashi M, Matsumoto S (2008) Contribution of activated interleukin receptors in trigeminal ganglion neurons to hyperalgesia via satellite glial interleukin-1β paracrine mechanism. Brain Behav Immun 22:1016–1023

Takeda M, Takahashi M, Nasu M, Matsumoto S (2011) Peripheral inflammation suppresses inward rectifying potassium currents of satellite glial cells in the trigeminal ganglia. Pain 152:2147–2156

Takeda M, Nasu M, Kanazawa T, Shimazu Y (2015) Activation of $GABA_B$ receptors potentiates inward rectifying potassium currents in satellite glial cells from rat trigeminal ganglia: *in vivo* patch-clamp analysis. Neuroscience 288:51–58

Tata AM, Vilaró MT, Agrati C, Biagioni S, Mengod G, Augusti-Tocco G (1999) Expression of muscarinic m2 receptor mRNA in dorsal root ganglia of neonatal rat. Brain Res 824:63–70
Tennyson VM (1965) Electron microscopic study of the developing neuroblast of the dorsal root ganglion of the rabbit embryo. J Comp Neurol 124:267–318
Tennyson VM (1970) The fine structure of the developing nervous system. In: Himwich WA (ed) Developmental neurobiology. Thomas, Springfield, pp 47–116
Thippeswamy T, Mckay JS, Morris R, Quinn J, Wong L-F, Murphy D (2005) Glial-mediated neuroprotection: evidence for the protective role of the NO-cGMP pathway via neuron-glial communication in the peripheral nervous system. Glia 49:197–210
Tischner K, Fischer HA (1975) Uptake of tritium labelled chloroquine into organized cultures of rat spinal ganglia. An electron microscope autoradiographic study. Acta Neuropathol 32:353–357
Tischner K, Murray MR (1972) The effects of sodium azide on cultures of peripheral nervous system. A. Light and electron microscope study. J Neuropathol Exp Neurol 31:393–410
Tischner K, Schröder JM (1972) The effects of cadmium chloride on organotypic cultures of rat sensory ganglia. A light and electron microscope study. J Neurol Sci 16:383–399
Truex RC (1939) Observations on the chicken Gasserian ganglion with special reference to the bipolar neurons. J Comp Neurol 71:473–486
Tuchweber B, Kovacs K, Khandekar JD, Garg BD (1972) Intramitochondrial lamellar formations induced by pregnenolone-16-α-carbonitrile in the hepatocytes of pregnant rats. J Ultrastruct Res 39:456–464
Unsicker K (1967) Über die Ganglienzellen im Nebennierenmark des Goldhamsters (Mesocricetus auratus). Ein Beitrag zur Frage der peripheren Neurosekretion. Z Zellforsch 76:187–219
Vaegter CB (2014) Neurotrophins and their receptors in satellite glial cells following nerve injury. Neural Regen Res 9:2038–2039
Valentin G (1836) Über den Verlauf und die letzten Enden der Nerven. Verhandlungen der Kaiserlichen Leopoldinisch-Carolinischen Akademie der Naturforscher 18:51–240
Valentin G (1839) Über die Scheiden der Ganglienkugeln und deren Fortsetzungen. Arch Anat Physiol Wiss Med 139–164
van den Bosch de Aguilar P, Vanneste J (1983) The microenvironment of the spinal ganglion neuron in the rat during aging. Exp Neurol 81:294–307
Van Gehuchten A (1892) Nouvelles recherches sur les ganglions cérébro-spinaux. La Cellule 8:233–253
Varon S (1976) Glia, nerve growth factor and ganglionic metabolism. In: Ahtee L (ed) Neurotransmission. Pergamon, Oxford, pp 275–284
Varon S, Raiborn C (1972) Dissociation, fractionation and culture of chick embryo sympathetic ganglionic cells. J Neurocytol 1:211–221
Varon S, Raiborn C, Tyszka E (1973) In vitro studies of dissociated cells from newborn mouse dorsal root ganglia. Brain Res 54:51–63
Varon S, Raiborn C, Burnham P (1974) Comparative effects of nerve growth factor and ganglionic nonneuronal cells on purified mouse ganglionic neurons in culture. J Neurobiol 5:355–371
Vause CV, Durham PL (2010) Calcitonin gene-related peptide differentially regulates gene and protein expression in trigeminal glia cells: findings from array analysis. Neurosci Lett 473:163–167
Vega JA, Rodriguez C, Medina M, del Valle-Soto ME, Hernandez LC (1989) Expression of cytoskeletal proteins in glial cells of dorsal root ganglia. Cell Mol Biol 35:635–641
Verbavatz J-M, Ma T, Gobin R, Verkman AS (1997) Absence of orthogonal arrays in kidney, brain and muscle from transgenic knockout mice lacking water channel aquaporin-4. J Cell Sci 110:2855–2860
Vesin M-F, Urade Y, Hayaishi O, Droz B (1995) Neuronal and glial prostaglandin D synthase isozymes in chick dorsal root ganglia: a light and electron microscopic immunocytochemical study. J Neurosci 15:470–476
Vit J-P, Jasmin L, Bhargava A, Ohara PT (2006) Satellite glial cells in the trigeminal ganglion as a determinant of orofacial neuropathic pain. Neuron Glia Biol 2:247–257

Vit J-P, Ohara PT, Bhargava A, Kelley K, Jasmin L (2008) Silencing the Kir4.1 potassium channel subunit in satellite glial cells of the rat trigeminal ganglion results in pain-like behavior in the absence of nerve injury. J Neurosci 28:4161–4171

Wagner R (1846) Sympathischer Nerv, Ganglienstructur und Nervenendigungen. In: Wagner R (Hrsg) Handwörterbuch der Physiologie 3/I. Vieweg, Braunschweig, S360–406

Wakisaka H, Kobayashi N, Mominoki K, Saito S, Honda N, Hato N, Gyo K, Matsuda S (2001) Herpes simplex virus in the vestibular ganglion and the geniculate ganglion—role of loose myelin. J Neurocytol 30:685–693

Wall PD, Devor M (1983) Sensory afferent impulses originate from dorsal root ganglia as well as from the periphery in normal and nerve injured rats. Pain 17:321–339

Watanabe PG, Sharma RP (1975) Neurotoxicity of organophosphates. Effects of tri-o-tolyl phosphate in chick ganglia cell cultures. J Comp Pathol 85:373–381

Waxman SG, Dichter MA, Hartwieg EA, Matheson JK (1977) Recapitulation of normal neuroglial relation in dissociated cell cultures of dorsal root ganglia. Brain Res 122:344–350

Weick M, Cherkas PS, Härtig W, Pannicke T, Uckermann O, Bringmann A, Tal M, Reichenbach A, Hanani M (2003) P2 receptors in satellite glial cells in trigeminal ganglia of mice. Neuroscience 120:969–977

Weis P (1971) The in vitro effect of the nerve growth factor on chick embryo spinal ganglia: an electron microscopic evaluation. J Comp Neurol 141:117–132

Wen JYM, Morshead CM, van der Kooy D (1994) Satellite cell proliferation in the adult rat trigeminal ganglion results from the release of a mitogenic protein from explanted sensory neurons. J Cell Biol 124:1005–1015

Werner MH, Nanney LB, Stoscheck CM, King LE (1988) Localization of immunoreactive epidermal growth factor receptors in human nervous system. J Histochem Cytochem 36:81–86

Wetmore C, Olson L (1995) Neuronal and nonneuronal expression of neurotrophins and their receptors in sensory and sympathetic ganglia suggest new intercellular trophic interactions. J Comp Neurol 353:143–159

Whetsell WO Jr, Bunge RP (1969) Reversible alterations in the Golgi complex of cultured neurons treated with an inhibitor of active Na and K transport. J Cell Biol 42:490–500

Whetsell WO Jr, Schwartz J, Elizan TS (1977) Comparative effects of herpes simplex virus types 1 and 2 in organotypic cultures of mouse dorsal root ganglion. J Neuropathol Exp Neurol 36:547–560

Wilkinson R, Leaver C, Simmons A, Pereira RA (1999) Restricted replication of herpes simplex virus in satellite glial cell cultures clonally derived from adult mice. J Neuro Virol 5:384–391

Woodham P, Anderson PN, Nadim W, Turmaine M (1989) Satellite cells surrounding axotomised rat dorsal root ganglion cells increase expression of a GFAP-like protein. Neurosci Lett 98:8–12

Woodhoo A, Dean CH, Droggiti A, Mirsky R, Jessen KR (2004) The trunk neural crest and its early glial derivatives: a study of survival responses, developmental schedules and autocrine mechanisms. Mol Cell Neurosci 25:30–41

Wu H-H, Bellmunt E, Scheib JL, Venegas V, Burkert C, Reichardt LF, Zhou Z, Fariñas I, Carter BD (2009) Glial precursors clear sensory neuron corpses during development via Jedi-1, an engulfment receptor. Nat Neurosci 12:1534–1541

Wyburn GM (1958) The capsule of spinal ganglion cells. J Anat 92:528–533

Xian CJ, Zhou X-F (1999) Neuronal-glial differential expression of TGF-α and its receptor in the dorsal root ganglia in response to sciatic nerve lesion. Exp Neurol 157:317–326

Yamadori T (1970) A light and electron microscopic study on the postnatal development of spinal ganglia in rats. Acta Anat Nippon 45:191–205

Yamamoto M, Fan L, Wakayama T, Amano O, Iseki S (2001) Constitutive expression of the 27-kDa heat-shock protein in neurons and satellite cells in the peripheral nervous system of the rat. Anat Rec 262:213–220

Yamashita N, Sakai K, Furuya S, Watanabe M (2003) Selective expression of L-serine synthetic enzyme 3PGDH in Schwann cells, perineuronal glia, and endoneurial fibroblasts along rat sciatic nerves and its upregulation after crush injury. Arch Histol Cytol 66:429–436

Yarygin KN, Doronin PP, Rodionov IM, Yarygin VN, Giber LM (1976) The study of neurons and glial cells in sympathetic ganglia in mice of different age. III. A study of perikaryal satellite cells in normal animals and in mice with sharply reduced number of ganglionic neurons (Russian text). Tsitologiia 18:944–949

Yates RD (1961) A study of cell division in chick embryonic ganglia. J Exp Zool 147:167–181

Yntema CL (1937) An experimental study of the origin of the cells which constitute the VIIth and VIIIth cranial ganglia and nerves in the embryo of Amblystoma punctatum. J Exp Zool 75:75–101

Yntema CL (1943) An experimental study on the origin of the sensory neurones and sheath cells of the IXth and Xth cranial nerves in Amblystoma punctatum. J Exp Zool 92:93–119

Yonezawa T, Iwanami H (1966) An experimental study of thiamine deficiency in nervous tissue, using tissue culture technics. J Neuropathol Exp Neurol 25:362–372

Yonezawa T, Mori T, Nakatani Y (1969) Effects of pyridoxine deficiency in nervous tissue maintained in vitro. Ann N Y Acad Sci 166:146–157

Young JAC, Brown DA, Kelly JS, Schon F (1973) Autoradiographic localization of sites of [^{3}H] γ aminobutyric acid accumulation in peripheral autonomic ganglia. Brain Res 63:479–486

Zerboni L, Arvin A (2015) Neuronal subtype and satellite cell tropism are determinants of varicella-zoster virus virulence in human dorsal root ganglia xenografts in vivo. PLoS Pathog 11:e1004989

Zhang J-M, Donnelly DF, Song X-J, LaMotte RH (1997) Axotomy increases the excitability of dorsal root ganglion cells with unmyelinated axons. J Neurophysiol 78:2790–2794

Zhang Y, Roslan R, Lang D, Schachner M, Lieberman AR, Anderson PN (2000) Expression of CHL1 and L1 by neurons and glia following sciatic nerve and dorsal root injury. Mol Cell Neurosci 16:71–86

Zhang X-F, Han P, Faltynek CR, Jarvis MF, Shieh C-C (2005) Functional expression of $P2X_7$ receptors in non-neuronal cells of rat dorsal root ganglia. Brain Res 1052:63–70

Zhang X, Chen Y, Wang C, Huang L-YM (2007) Neuronal somatic ATP release triggers neuron-satellite glial cell communication in dorsal root ganglia. Proc Natl Acad Sci U S A 104:9864–9869

Zhang H, Mei X, Zhang P, Ma C, White FA, Donnelly DF, LaMotte RH (2009) Altered functional properties of satellite glial cells in compressed spinal ganglia. Glia 57:1588–1599

Zhou X-F, Rush RA, McLachlan EM (1996) Differential expression of the p75 nerve growth factor receptor in glia and neurons of the rat dorsal root ganglia after peripheral nerve transection. J Neurosci 16:2901–2911

Zhou X-F, Deng Y-S, Chie E, Xue Q, Zhong J-H, McLachlan EM, Rush RA, Xian CJ (1999) Satellite-cell-derived nerve growth factor and neurotrophin-3 are involved in noradrenergic sprouting in the dorsal root ganglia following peripheral nerve injury in the rat. Eur J Neurosci 11:1711–1722

Ziegler RJ, Herman RE (1980) Peripheral infection in culture of rat sensory neurons by herpes simplex virus. Infect Immun 28:620–623

Ziegler RJ, Pozos RS (1977) Ultrastructural effects of herpes simplex virus type 2 infection of rat dorsal root ganglia in culture. J Neuropathol Exp Neurol 36:680–692

Zimmerman E, Karsh D, Humbertson A Jr (1971) Initiating factors in perineuronal cell hyperplasia associated with chromatolytic neurons. Z Zellforsch 114:73–82

Zimmermann A, Sutter A (1983) β-Nerve growth factor (βNGF) receptors on glial cells. Cell-cell interaction between neurones and Schwann cells in cultures of chick sensory ganglia. EMBO J 2:879–885

The manufacturer's authorised representative in the EU is Springer Nature Customer Service Centre GmbH, Europaplatz 3, 69115 Heidelberg, Germany. If you have any concerns regarding our products, please contact ProductSafety@springernature.com

Printed and bound by CPI Group (UK) Ltd, Croydon, CR0 4YY
29/09/2026
02224904-0001